SUSTAINABLE DEVELOPMENT
Issues and Case Studies

About the Authors

Srikumar Chattopadhyay (*b.* 1953) obtained M.Sc. (1974) and Ph.D. (1986) degrees in Geography from the University of Calcutta. Presently, he is working as a scientist in the Centre for Earth Science Studies, Thiruvananthapuram.

His research interest covers sustainable development, environment-development interface, microlevel planning, applied geomorphology and participatory resource evaluation.

He has contributed significantly in developing Participatory Resource Mapping model (PRM) now widely used in the country for watershed management and participatory planning.

Dr. Chattopadhyay is the Chairman of NAGI Commission on Sustainable Development.

Professor **H.S. Sharma** (*b.* 1945) holds postgraduate and doctoral degrees in geography from the University of Saugar, Madhya Pradesh.

He has been participating in several national and international conferences such as First, Second and Third International Conferences on Geomorphology held at Manchester, UK (1985) Frankfurt, FRG (1989) and Hamilton (1993), the Desert Sediment meeting, London (1986) and International Symposium on Aerospace Survey and Natural Disaster Reduction, Enschede, Netherlands (1989). Professor Sharma has also attended first international Science Conference under the auspices of International Geosphere-Biosphere Programme (1995) at Garmisch, Germany, and visited Salford and East Anglia (Norwich) Universities under Academic Link Exchange Programme (1993) and Norway (1995).

Currently Dr. Sharma is a Professor and Head of the Department of Geography, University of Rajasthan, Jaipur, and also working on a Project on Ecoperception of Ranthambhore Sanctuary sponsored by Ministry of Environment and Forests, New Delhi.

SUSTAINABLE DEVELOPMENT
Issues and Case Studies

Editors
SRIKUMAR CHATTOPADHYAY
H.S. SHARMA

Published for
Commission on Sustainable Development
National Association of Geographers, India (NAGI)

By

CONCEPT PUBLISHING COMPANY, NEW DELHI-110 059

ISBN 81-7022-694-5

First Published 1998

Published and Printed by
Ashok Kumar Mittal
Concept Publishing Company
A/15-16, Commercial Block, Mohan Garden
NEW DELHI-110 059 (India)

Phones : 564 8039, 564 9024
Fax : 09-11-564 8053

FOREWORD

People who are happy with the way humanity has been managing its affairs for the last two centuries are not many. It is widely accepted that the goal of continuing economic growth seen in global perspectives can be only by jeopardising the very survival of life on planet earth. This realization is, however, still at the awareness level; when it comes to action much of the business of development is as usual; it has not changed much.

If awareness is the first step in the process of action, then there is a ray of hope. Humanity has to move fast on a sustainable development trajectory. The task is, however, difficult, if not impossible. The bug of rapid economic growth, unmindful of what happens to the environment and human values, has infested all spheres of human activity, including our personal life, family, community, society and the whole earth. It has changed the conditions of life and human relations to an extent that a completely new world view for evaluating the meaning and purpose of life has gripped the minds of the people.

No matter which part of the world one visits and which individual or group of people one meets, one gets struck by this new world view centred on material avarice and greed, side living, may throwing overboard the values which endangered love, fellow feeling, concern about others both living and non-living. True, there were wars and conflicts in the past, but even these had rules and were limited to a small section of the ruling class. It is also true that under the old dispensation, the poor suffered from not one but many disabilities, but given the scientific and technological base of that society, its effects and impacts were limited to only the local spheres.

The weapons of war were simple and could be possessed and used by the brave; the tools of work were also simple, and hence could not become instruments of mass destruction of nature to give way for agriculture or to promote industry. The social solidarity perhaps borne out of survival instinct was so strong that much of what was natural was part of commons property. It could not be owned by individuals. And hence the life forces in human beings ran parallel to those in the natural environment. Contradictions if any, were within the capacity of the nature to absorb and clear.

The great changes that agricultural revolution ushered in, thousands of year ago, were grow and hence there was enough time for man and the nature to understand both the short and long term consequences of imbalances so brought about and make necessary adjustments so that the work of the nature and that of man are not counterposed, rather, they become complementary. Even then, the aberrations were not uncommon and the rape of the earth was one of the major causes of the down fall of some of the great civilizations of the past.

Then comes the second wave of change popularly known as Industrial Revolution. It strengthened the hands and minds of men first in England and Wales, then in Germany beyond imagination. Man was taught to control nature and ultimately use it to improve his material well-being. Nature become a resource for exploitation—a resource shorn of life support elements. Forest become timber; and become real estate; wild animals become game; and minor animals, insects and micro-organisms become enemies of man and hence of targets of insecticides and pesticides. The symbiosis of nature and food chain that maintained it was disturbed and at places destroyed.

The last two decades of the 20th century saw a growing concern for the widening gap between human greed and nature's capacity to reproduce itself to satisfy this greed. Alarms have been raised all over the world about ozone layer depletion, climatic changes, pollution of the air, water, and land, all caused by man and his activities. The rate at which bio-diversity is being diploid, has put a question mark on the very survival of life on the earth. Coming down to the regional and national level, the situation has become highly disturbing. Both the rich and the poor of the world are undermining the natural resource base of the earth in wanton manner. The top 20 per cent of the world population uses 75 per cent of earth resources, while the bottom 20 per cent lives on meagre 5 per cent. Both the rich and the poor are overdrawing on resources one because of overconsumption and the other because of a huge population mass.

The old and time tested values which maintained a symbiotic relationship between man and nature is gone and a new consumerist culture has taken deep roots. Nature is no longer a living phenomenon; it has become a bundle of commodities to be used to meet the evergrowing wants of man.

The concept of sustainable development is a first major ideational breakthrough to halt this rape of the nature. It is ideational because it has moulded to a great extent the thinking of the people all over the world. The idea underlying the concept was already there underneath the occasional protests, writings and warnings given by statesmen and humanists from time to time. It came to the surface in 1980s with the publication of the Brundtland Report.

I said that sustainable development is only the first step. It is the first step because it does not attack the root cause of the problem. The root cause is our materialistic world view. Sustainable development is not able to divorce itself from development which essentially means material development. In fact, development in its present form is the root cause of environment destruction. Sustainable development only aims at limiting or reorienting it to sustainability. The emphasis is on continuing development but with lesser use of natural resources and lesser pollution. The right course would be modify the meaning and style of development, so that nature's use is limited to its capacity of recoup. In other words, the scope of development should not be confined to material achievements alone; it should incorporate healthy body and mind. As a well developed person will be loving, ever smiling and compassionate, a developed society too will be full of human brotherhood, strong integrated communities and happy families. Material well-being is of course important to achieve all these qualities but material well-being alone will ensure these.

Already there are new measurements of development to include health, education etc. in addition to income. They, however, exclude style of life, the world view people adhere to, and values. In fact beyond a certain level of material welfare, it is the non-material gains that enrich life and make people happy. This is what sustainable development does not address to.

Yet in a world of abject materialism, sustainable development is a good starting point. The tragedy is that even this little step forward is being shelved or sidetracked by most of the countries of the world—developed as well as underdeveloped, or one pretext or the other. One of the reasons for slow progress in this direction is the pressure of the development paradigm on the lives of the people and lack of understanding about what sustainable development involves.

Several publications have come out in recent years on sustainable development but there is a need for more literature dwelling on the theory and practice of sustainability. I am glad Professor H.S. Sharma and Dr. Srikumar Chattopadhyay have brought out an excellent volume on this theme. The volume owes its origin to the XIV Indian Geography Congress held at Jaipur in 1992. A special session on sustainable development was organized under the auspices of the National Association of Geographers, India (NAGI) Commission on Sustainable Development. The session was so well attended and some of the papers presented were so good, that it was decided to publish selected papers after the Congress. Prof. H.S. Sharma and Dr. Srikumar Chattopadhyay were given the responsibility to edit the volume.

I am glad that efforts of Prof. H.S. Sharma and Dr. Srikumar Chattopadhyay have borne fruit and a very useful volume has been edited. It includes some of the papers discussed at the Jaipur Congress. To fill the gaps, they have included some invited papers of relevance.

I congratulate the editors for the excellent work done and that the book will be read widely and will find a place in all the libraries of India and abroad.

R.P. MISRA
President,
National Association of Geographers
India

December 1, 1997

ACKNOWLEDGEMENTS

We, the members of the commission on Sustainable Development, National Association of Geographers, India (NAGI) are thankful to the Executive Committee (EC) of the National Association of Geographers, India for giving us this opportunity of organising the Commission activities.

We are grateful to Prof. R.P. Misra, former Professor in Geography, Department of Geography, Delhi University and former Director, Gandhi Bhavan, Delhi University under whose initiative this work could be undertaken. He has taken personal interest in guiding us and making arrangement for publication. Prof. B. Thakur, Secretary General, NAGI is thanked for taking special interest in commission activities.

We express our deep sense of gratitude to all the contributors for their ready response to our request for sending the articles and bear with us the inevitable delay to bring out this volume.

We are grateful to the Director, Centre for Earth Science Studies,Trivandrum for providing facilities to undertake this work. Sri K. Surendran has assisted in secretarial activities. Thanks are due to the Concept Publishing Co. for publishing this book.

INTRODUCTION

Sustainable development is perhaps the most quoted word in past one decade, specially after publication of Our Common Future—A Report of the World Commission on Environment and Development in 1987. The Earth Summit at Rio in 1992 debated the issues covering environment development interrelationship and a world wide programme of action had been chalked out under the caption Agenda 21.

People from all walks of life involved directly or indirectly in development planning is convinced that some definite directional change in planning processes has to be initiated, laissez faire affair can no longer go on.

It is now well realised that multi disciplinary, local level assessment like approach has to be followed to deal with this peculiar situation where resource utilisation has to be optimised with minimum injury to the environment..

Considering overriding importance of the issue and its societal implication, the National Association of Geographers, India (NAGI) appointed a Commission on Sustainable Development. The first meeting of the Commission took place in 1992 at Jaipur. It was decided that one of the activities of the Commission will be to look into the various issues involved in this matter from a multidisciplinary angle. Accordingly, it is decided that a volume incorporating invited articles will be brought out as Commission report. Professionals and academicians of various disciplines, both in India and abroad, were approached to contribute towards this volume.

Altogether 22 articles are chosen for this book. The articles are arranged into two sections, namely, (i) Issues and Approach; and (ii) Case Studies and Experiences. There are 10 articles in Section I and the Section II contains 12 articles. Attempts are made to follow a logical sequence while arranging the articles.

Along with concepts and issues the articles in the first section also provide methodological details citing definite examples. Prof. Misra's article has dealt with the ecological issues at length and indicates a definite course of action. Dr. Parameswaran argued that environmental issues have to be considered as part and parcel of the development planning. Many a problems arise due to narrow conceptual understanding of the environment development relationship. A very practical problem of quantitative measurement of sustainability is addressed by Dr. Carpenter.

The article on Politics and Economics of Agenda 21, Rio Summit by Dr. Thomas Isaac raised certain basic issues. The author argues that environmental problems can not be isolated from the political economy and the mode of production process operative in the World today. Hema Yadav in her article draws our attention to politicisation of environmental problems at global level. Dr. Chattopadhyay presents a broad review of the sustainable development concept and identifies some definite course of action with a methodological outline.

The article on Linkage Between Poverty and Environmental Degradation: A Profound Opinion by Dr. Fidel ties to draw our attention to some burning problems faced by the developing nations.

How ecosystem approach will help in sustainable rural development has been discussed by Dr. K.V. Sundaram with concrete examples. Dr. Anjana Desai dealt with theoretical aspects of resource evaluation for rural development.

Community management is one of the important issues in sustainable development. An in-depth analysis is presented by Dr. John Kurian in the context of coastal fisheries.

The case studies cover a wide spectrum across the country. Most of the studies refer to the micro level. The article by Sri T. Madhav Menon brings out importance of traditional knowledge in achieving sustainable development. His study is in the context of tribal society in Kerala. Prof. Chattopadhyay and his group present the case of a tribal community in West Bengal. These two studies, one in Kerala and another in West Bengal call for looking to the tribal society as repository of traditional knowledge.

Sustainable Agricultural Development has been discussed by Dr. B.L. Sharma and Palak Bhardwaj in a broader context and also for Rajasthan. A down to earth field experience is reported by Prof. Ghatak for a village in West Bengal. Farm system research is one of the significant developments took place during the last 2 decades in the field of agricultural sciences. As a tool of sustainable agricultural development this method can contribute to a large extent.

Study of Dr. Mridula Niyogi stresses on water resource management for sustainable development. With concrete examples from Ajay river basin of West Bengal she has successfully brought out a water resource management framework.

Smt. Mahamaya Chattopadhyay presents a method of landform analysis for sustainable landuse development. Although it is location—specific the method is replicable to cover undulating hilly region. Smt. Sunita Purakayastha has attempted to advocate a method of village level analysis for waste land development. This is quite significant in the Maharashtra state, where wasteland management is a problem.

Ecologic—economic analysis of a panchayat has been attempted by Sri C.K. Sasidharan. The approach is useful for panchayat level planning. Prof. B. Eswarappa has brought out importance of Common Good Traditions in village life. His argument is that loss of traditional values is a matter of concern in planning for sustainable society.

Dr. K. Soman, in his paper, dealt with environmental impact of mining activities in Kerala. Mining is one of the major activities that has far reaching impact on environment. EIA of mining is necessary for planning sustainable development.

A case study on Philippines by Smt. Elnora R. Pulma highlights problems of over consumption in a developing Asian nation. Dr. B.K. Roy's paper on Population and Development: Some Issues—India's profile, deals with demographic characteristics of the country. It also presents certain guidelines for development considering population variable.

Through these 22 articles a broad spectrum is covered to provide clear understanding of the complex nature of the problems related to sustainable development. Multidisciplinary view point expressed in these articles will help the readers to a great extent.

We do hope that this volume will be useful to the professionals, academicians, research scholars and the students of geography, economics, planning and other allied disciplines.

Jaipur H.S. SHARMA
Thiruvananthapuram SRIKUMAR CHATTOPADHYAY

CONTENTS

PART II : CASE STUDIES AND EXPERIENCES

LIST OF CONTRIBUTORS

Dr. Anjana Desai, Reader in Geography, Department of Geography, Gujarat University, Ahmedabad

Dr. B.K. Roy, Former Director, National Atlas and Thematic Mapping Organisation, Government of India, Calcutta

Dr. B. Eswarappa, Assistant Professor in Geography, Department of Geography, Bangalore University, Bangalore 560 056

Dr. B.L. Sharma, Former Head, P.G. Department of Geography, Government College, Beawar

C.K. Sasidharan, Technical Officer, Centre for Earth Science Studies, Thiruvananthapuram, 695 031

Chandreyye Das, Rural Development Centre, Indian Institute of Technology, Kharagpur 721 302

D. Bhattacharya, Department of Humanities and Social Sciences, Indian Institute of Technology, Kharagpur 721 302

Elnora R. Pulma, Planning Officer, Commission on Population, Government Centre, Candahug Pale, Layte, Philippines

Dr. Fidel Marquez Sanchez, Professor in Economics, Department of Economic Development, Economics Institute, University of Havana, Cuba

Hema Yadav, Lecturer, Department of Geography, Government Girls' College, Ajmer

Dr. John Kurian, Associate Fellow, Centre for Development Studies, Prasantnagar, Thiruvananthapuram, 695 011

Dr. K.V. Sundaram, FAO Consultant, Vasantkunj, New Delhi

Dr. M.P. Parameswaran, President, All India People's science Network, Bharat Gyan Vigyan Samithi, R.K. Puram, New Delhi

Mahamaya Chattopadhyay, Technical Officer, Centre for Earth Science Studies, Thiruvananthapuram, 695 031

Dr. Mridula Niyogi, Lecturer in Geography, Shibnath Sastri College, Calcutta

Palak Bhardwaj, Co-Editor, Bhoodarshan, Udaipur

Dr. R.N. Chattopadhyay, Professor, Rural Development Centre, Indian Institute of Technology, Kharagpur 721 302

Dr. Richard A. Carpenter, Former Associate Fellow, East West Environment and Policy Institute, Hawaii, USA

Dr. R.P. Misra, Director, Gandhi Bhavan, Delhi University, Delhi

Dr. S.S. Ghatak, Reader and Head, Department of Agricultural Entomology, Bidhan Chandra Krishi Vidyalaya, Kalyani, 741 252

Dr. K. Soman, Scientist E1 & Head, Resource Analysis Division, Centre for Earth Science Studies, Thiruvananthapuram, 695 031

Dr. Srikumar Chattopadhyay, Scientist E1, Centre for Earth Science Studies, Thiruvananthapuram, 695 031

Sunita Purakayastha, Research Fellow, Lake Gardens, Calcutta 700 045

Tapati Dasgupta, Rural Development Centre, Indian Institute of Technology, Kharagpur, 721 302

T. Madhav Menon, Editor, International School of Dravidian Linguistics, Thiruvananthapuram, 695 586

Dr. T.M. Thomas Issac, Associate Fellow, Centre for Development Studies, Thiruvananthapuram, 695 011

LIST OF TABLES

LIST OF FIGURES AND MAPS

PART I

ISSUES AND APPROACH

1

SUSTAINABLE DEVELOPMENT
The Ecological Perspectives

R.P. MISRA

Introduction

THE concept of Sustainable Development emerged as an important theme in 1987. The Brundtland Commission on Environment and Development in its report Our Common Future called for application of sustainability as a criterion for all development initiatives. "In order for development to be sustainable, it must meet the needs of the present without compromising the ability of future generations to meet their own needs", said the Commission.

This definition is highly generalized and hence has been subjected to a variety of interpretations. Broadly speaking, developed countries lay emphasis on economic and technological aspects. How to integrate environmental costs into economic decision-making while protecting and enhancing their industrial competitiveness and employment opportunities. The general feeling is that the current development/growth processes can continue provided the technological innovations are rapid and appropriate enough to reduce environmental side effects.

In contrast, the perspectives of developing countries are understandably different. In their view, poverty is the greatest polluter of environment and hence they must accelerate economic growth to meet the basic needs of the people. And if the industrialized rich countries are really serious about environment, they should realise that it is they who have created the present environment crisis, not the poor countries, and hence they should assist the latter with finance and latest environment friendly technological knowhow to promote economic growth.

Interestingly, both sides have a firm faith in economic growth and all that goes with it, such as industrialisation, urbanisation, and environmental deterioration. Both pin their hopes in science and technology to solve the emerging problems. It is simply the question of access to resources, particularly finances and technology. One wants to maintain the status quo, i.e., its economic and technological superiority and competitiveness; the other wants to change the existing world order to make it more egalitarian. The means for achieving the goals are the same.

Defining Sustainability

In view of these divergent approaches, it is almost impossible to define sustainable development more sustainably than done by Brundtland Report. Each country and each region assigns different weights to different approaches to achieve a sustainable future. Each definition is interest-oriented and genuine efforts to look at the issue from the point of view of human survival and development have remained confined to a limited number of scholars and thinkers.

Sustainable development cannot be a static concept and hence cannot be defined once for all. It is a dynamic process and will be applied by different countries in tune with their own cultural, political and economic perspectives. But unless there is a broad agreement on goals, directions and means, the concept will prove sterile and the business of development will continue as usual with more tensions among the nation states and regions and localities within the nation state.

An attempt is, therefore, being made here to offer a broad outline of what constitutes sustainable development or how sustainable development differs from development. In general (i) it cannot be achieved in the short run; (ii) it is based on equity and justice; (iii) its approach is balanced and integrative; (iv) it has common goals but different routes; (v) it accepts nature not only for as a resource for development but also as the earthly womb for survival and development of humankind; and (vi) it is participatory in nature.

Sustainable development has to be a long-term process, and one may say a continuing process. The economic, political, technological and social structure and superstructure built during the last few centuries the world over, particularly in industrialized countries of the world, cannot be dismantled in a day. No structure is, however, permanent. Elemental and partial changes do take place as a part of renewal process. This renewal process can be accelerated within the tolerable birth pangs and death agonies that mark any such change. The renewal process must aim to achieve the characteristics of sustainable development listed above.

Equity and justice are the *sin qua non* of sustainable development. It is four dimensional: (1) Equity among nations—rich nations, poor nations, developed nations, developing nations; (2) Equity within countries— between regions, social classes, genders, sectors of activities; (3) Equity between generations; and (4) Equity between economics and ecology and science and spirituality.

The first two equities, although obvious, yet need some explanation. Let it not be construed to mean that pollution, for example, should be equally distributed. Equity is not equality; nor does it have any negative connotations. Equity promotes variety but not disparity. It is positive in the sense that it does call for a fair distribution of harmful things. Moreover, it is directly linked with ability and capacity to bear the burden. This means that rich countries, rich regions, rich groups and the favoured sex will have to take greater responsibility for sustainability.

Equity between generations invariably raises contentious issues like: Why should the present generation bother about the future? How can we be sure that future generations would not be better off in terms of scientific knowledge and technological know-how to resolve their problems? Presuming that the dependence of the future generations on nature will not be less than our own, humanity demands that we should not defer costs of present development to our children. Is there any doubt that the future generations too would need fertile soil, scenic beauty, clean air, potable water, virgin forests, energy, and wild life?

Lastly, we must try to bring about equity between our economic and ecological interests. The last few centuries have seen economic goals being pursued is unmindful of what is going to happen to the Earth which sustains life. We are witnessing and experiencing the consequences of this approach. Similarly, science and bounded rationality have guided our thoughts and actions. Today, its social and environmental consequences are obvious to us. Economics and science have great potentials for human welfare. In fact, they have taken humanity to a point of achievement beyond imagination. But their side-effects on man and environment are proving to be unbearable. What we need now is the integration of economics and ecology, and science and spirituality to cure the developmental disease inflicting humanity and also to carry man safely to new heights of glory.

Goals of Sustainable Development

All religions have a common goal, i.e., realization of the self. They follow different paths to arrive at this common point. If all of them had employed right means, world would have been happier today.

Sustainable development too has a fixed set of goals. Approaches and means to achieve them can differ. These goals are: (1) Basic needs of all human beings, i.e., food, clothing, shelter, education, health, security, and self-esteem must be met adequately. Priority must go to these needs. Level of these needs will be determined by natural and technological resources available and the global socio-economic context; (2) Development processes should be so articulated that ecological balance and environmental purity is least disturbed, if at all; and (3) All nations and people must join hands to support each other and work with each other to create a world in which the above two goals are optimised. Each country should find ways and means to promote this interdependence.

Sustainable development demands a change in our attitude towards nature. Nature is not just a material resource for human consumption. Man is also a part of nature. What will happen if a baby in the mother's womb acquires the power to break open the womb and come out prior to the time of its maturity? Both the mother and baby will be in trouble. There is an urgent need, therefore, to move away from the Cartesian world-view and post ourselves in the ecological world-view. The idea is not to go back to the past, but to create a new future out of the present, the present as it is. The past can help in the shaping of the new future.

And finally, sustainable development is not the business of the government and private companies alone. It is the business of the people in general. It is a process which has to be initiated at each level of human endeavour and life. It involves individuals, families, communities, corporate bodies, nations and global society. It has to be a movement, because it involves paradigmatic change which is difficult to bring about unless great many people get involved. Democratisation of decision-making and decentralisation of power and authority is, therefore, a must.

Thus, we see that sustainable development is a comprehensive term. It involve far more than environment friendly technology. It involves paradigm change; it involves change in style of development; it involves change in our attitude towards nature; and it involves political, social and ethical changes. No wonder then that it does not have a single definition on which all agree.

Sustainable development can be looked at from a variety of perspectives, such as environmental, economic, social, political, cultural and technological. None of the perspectives can stand alone. Sustainable development is possible, if all of them are taken into consideration. Their identification as such is designed to help specialists in various fields to find their own place in the task ahead. Such an approach will allow in-depth studies without forgetting the breath of the problem. In this article, environmental sustainability has been taken for deeper analysis.

Environmental Sustainability

Man-Nature Relationship

Environmental sustainability aims at ensuring the protection, conservation and better management of Earth's natural resources. The idea is not to debar people from using these resources. It rather aims at such a pattern of use, that resources, particularly those which are so vital for human survival, are neither exhausted nor polluted nor destroyed. This would require a change in our attitude towards Nature as also towards what we call economic development

Maurice F. Strong, secretary General of the UNCED 1992, said in 1989:

> "We have been literally living off the Earth's capital, and that capital is being seriously depleted. We cannot continue to run our planet this way anymore than we could run a business for long by running down and living off its capital. Indeed our Earth, if it were an incorporated entity, would be headed for bankruptcy if we continue on our present course."

Richard St. Barbe-Baker said the same thing more tellingly:

> "This generation may either be the last to exist in any semblance of civilized world or it will be the last to have the vision, the bearing and the greatness to say: "I will have nothing to do with this destruction of life; I will play no part in this devastation of land; I am determined to live and work for peaceful construction, for I am morally responsible for the world of today and the generations of tomorrow".

Let us now examine environmental sustainability in greater detail. It is rarely realised that human beings, like other living entities, have come from the Nature. Nature has a nurturing role. After death, the human body, the part that is matter, goes back to Nature. The Hindu concept of Panchabhoota is so relevant here. Land, water, air, energy, and space constitute the basic elements of all living organisms. All these must remain ample and pure to sustain life. The Vedic poet Atharva wrote a homage to Earth and incorporated it as 'Prithvi Sukta' of Atharva Veda. The essence of the Sukta is:

> The Earth is the mother and we are her children. People of other religious traditions may not be used to such a reverence for nature. But they cannot ignore the fact that the human destiny is inseparably linked with that of the Earth. It is, therefore, our duty to maintain the Nature of nature. It is in our self interest, if nothing else.

Ecosystem Integrated

Ecosystems take a long time to reach climax. The climax is marked by four characteristics. These are complexity, stability, diversity, and resilience. All natural ecosystems are characterized by these qualities. If any of these characteristics are weakened or jeopardised, the system collapses.

Ecosystems have great resilience; they recover from minor disturbances in no time. By minor, we mean within the limits of tolerance. Human body is also an ecosystem. Its metabolism removes the unwanted things. Even a minor doze of poison is washed off. But too much of poisoning through smoking, drinking, junk food, polluted air and water and what not, is beyond the capacity of the system to clean; it weaken its resilience.

Unfortunately, we are unable to see the long-term effects of over loading the body beyond certain threshold until it breaks down. Minor and temporary illnesses do not bother us because we have medicines for them. The toxic medicines further overload the system to invite cancer, heart attack, Aids, and so on. What is happening to the body system, is also happening to the ecosystem. More often than not, we do not realise what harm we have inflicted to it until it is too late. The ecosystem is destroyed.

The ecosystem integrity can be maintained if we know its carrying capacity, its assimilative capacity and its renewability. Carrying capacity is the ability of an ecosystem to feed a certain population of humans and/or animals. Use of any resource beyond its carrying capacity will threaten its stability. Technology can be of some help. But once its natural capacity to regenerate dies, even technology cannot be of much help. What can the technology do, once the soil is exhausted or a species is lost? It will be too late and costly to do anything.

In the case of land, ecologists have worked out the carrying capacity in terms of Net Primary Productivity (NPP). NPP can be calculated for each piece of land. Global analysis of NPP shows that it is declining very fast, because of deforestation and over-grazing. Attempts are being made to work out Net Water Productivity (NWP). It is estimated that the water released in irrigation canals of India can be easily reduced by 30 per cent, if it is not wasted. If the flow irrigation is managed in the same way as the traditional irrigation system (wells and tanks), the total requirement can be reduced by 50 per cent. And if drip irrigation is introduced, the amount will be further reduced. The surplus can then be utilised to irrigate tail-end lands in the command areas of canals, as also the parched land outside it. We have drawn water of some of our rivers beyond their carrying capacity. Yamuna in Delhi is nothing but a cess pool. We can bring our rivers back to life without any adverse effect on production and productivity.

The same applies to forests. Large tracts of our forest were destroyed because the felling of trees and grazing by animals went beyond their carrying capacity. Therefore, one has to ensure that population of human beings or animals does not go beyond the carrying capacity of the ecosystem.

Since the ecosystem is integrated, i.e., its constituent elements are interdependent, destruction of any of the constituents will harm the whole system. Land, soil, water, vegetation, temperature, pressure, animal life, including micro organisms constitute an integrated whole.

Assimilative capacity is another characteristic of the ecosystem. It is also directly linked to its carrying capacity. All systems, including the ecosystem, have a metabolic process to remove the waste. Certain amount of waste discharged in soil, air or water will be purified in the natural course. Problem arises if the amount of waste is beyond the natural capacity to clean or of a kind that cannot be handled by the ecosystem. In big cities, the amount of waste is too large to expect any river howsoever mighty in may be, to clean it. Waste, therefore, has to be treated before it is released or its quantity and quality have to be controlled. Similarly, some of the industrial waste cannot be handled by the ecosystem metabolism. Examples of mercury, acids and atomic wastes can be cited in this regard.

Recharging capacity of the ecosystem is its third characteristic. The example of shifting agriculture can be taken to bring home this point. The tribal farmer leaves the land to enable the soil to recoup fertility. And then after few years, he comes back. He knows how many years it would take for the land to recoup. All renewable resources need time to recoup. The impression that the renewable resources can be misused or mismanaged without any adverse effects, is totally wrong. What we need urgently as a policy measure is to see that renewable resources are not diverted to other uses. Secondly, we must restrict the waste emissions to conform to the assimilative capacity of the ecosystem.

Biodiversity

The third dimension of environmental sustainability is biodiversity which has to be seen from three perspectives-genetic diversity, species diversity and ecosystem diversity. Genetic diversity refers to the sum of chromosomal information carried by genes of individual plants and animals. If we destroy any plant or animal species, we would loose that gene for ever. Let us take the case of wheat. It has hundreds of varieties. Each variety is separate from the other in matters of gene. If everyone grows just a few varieties of hybrid wheat, the gene bank of wheat maintained by nature will be lost forever. The same applies to trees. If virgin forests are destroyed, one does not know how many varieties of plants and animals we would have lost. Many plants and animals live in association with each other and the destruction of any of them may lead to the destruction of associated life also.

We should be concerned about biological diversity because each of the plant and animal species is interdependent and contains gene pool of great importance. We do not yet know the contributions of these plants and animals to make this world a liveable place. According to Ayurveda, there is no plant (from moss to Peepal tree) which does not have some medicinal value. Same applies to animal life. We do not yet, know what will happen to remaining animal species, if those plants and animals on which they depend disappear.

In order to conserve the biodiversity, natural ecosystems as well as whatever is left of the modified ecosystems must be protected. Further, manmade ecosystems should now be planned, so that in the next 50 years or so, we have more such areas where nature has its own sway. Countries like India should be careful about clearing forests for diverting forest lands to agriculture. It is a very short-sighted policy.

Man-Land Ratio

The fourth aspect of environmental sustainability that needs our attention is population. "The quality of human life is inseparable from the quality of the environment. It is increasingly clear that both are inseparable from the question of human numbers and concentrations", says UNFPA(1990).

The carrying and assimilative capacities of the ecosystem depend on man-land ratio (land-nature). If human population increases without proportionate increase in natural resources, the carrying capacity of ecosystem, i.e., the quantity and quality of natural resources gets reduced. This problem is further accentuated, if the population increase takes place in those regions and nations where most of the people still depend on agriculture and other primary activities. As a part of their survival strategy, they cut trees and kill animals. When it is the question of hunger, no one can afford to think of the animals let alone the future. Poor people live a hand-to-mouth existence and on a day-to-day basis.

Movement of rural population to urban areas is another dimension having direct implications for environment and its sustainability. Even a cursory look at the slums of Indian metropolises tells more of what population concentration does to environment than anything else. In these cities, man is acquiring the traits of beasts. Isn't it that there is more wild life in Delhi today than in any of the wild life sanctuaries?

Environmental Planning

The last dimension of environmental sustainability brings us to environmental planning.

Its aim is four fold: (i) Environmental Protection; (ii) Rehabilitation and restoration of degraded ecosystems; (iii) Augmentation of the carrying capacity of both natural and human-managed ecosystems; and (iv) Creation, expansion and development of new ecosystems, both minor and major.

Tradition of Sustainable Environment in India

In this last section of the article, an attempt is made to have a glimpse of how India, prior to the colonial rule, was able to maintain sustainable environment. Sustainable environment was an in-built phenomenon ever since the Vedic times. The Cosmic Law, i.e., R'ta which kept order in the universe, is the same which keeps the ecosystem going. As the Rig Vedas say:

The Rivers flow R'ta. According to the R'ta, the light of the heaven born morning has come. . . . The year is the path of R'ta. The gods themselves are born of the R'ta or in the R'ta; they show by the acts that they know, observe and follow the R'ta. In man's activity, the R'ta manifests itself as the moral law.

To obtain natures bountry, man must obey R'ta. for one who lives according to Eternal Law, the winds are full of sweetness, the river pour sweet. So may the plants be full of sweetness for us.

This quotation from Rig Veda gives an idea of how an Indian looks at the universe and the ecosystem. These thoughts still pervade the Indian mind. How these thoughts were translated into reality? For this two examples are given below, both in historical context as well as contemporary context.

Conservation of Forests

The first example pertains to forests. There used to be three types of forests in India. Remnants of each type still exist in the country and they have their contemporary relevance also. These forests were biotic reserves as we have today but the way these have been managed was entirely different. It was then a participatory management by the people themselves. These were: (i) *Mahavana*, (ii) *Tapovana,* and (iii) *Srivana.*

Mahavana was a dense and natural forest of great expanse. Such forests were devoid of habitation and human interference was negligible, if at all. In these forests, biodiversity prevailed at its maximum. They were the main sources of medicines. Lord shiva was the presiding deity of *Mahavana*. No one encroached upon these forests. They remained virgin.

The *Tapovana*, as the very name indicates were the forests where sages did *tapas* (meditation) which had *ashrams* where *Gurus* taught the pupils and which were away from human habitation. People had access to these forests but only for peaceful purposes. It was in these forests that our *Upanishads* and *Aranyakas* were written. There are still some *Tapovanas* like the one near Haridwar.

As the *Tapovanas* were destroyed, the tradition of *rishis* and *sadhus*—living detached from the rest of the world and engaging themselves in meditation, writing, and teaching—also died. It is pertinent to note that the *Tapovanas* abounded in wild life but no one was allowed to kill animals. Even princes were punished for killing animals (as game) near the *Ashramas*. It was in these forests that the training in the use of weapons was also imparted. It was part of the overall training of pupils. Thus, the matters pertaining to both peace and war were really in the hands of saintly people. This is what ensured peace in the society. People dejected from life went to the *Tapovanas*, rather than committing suicide, as it happens today. The excommunicated criminals also went to these forests for repentance and cleansing themselves.

Srivanas were essentially the village or town forests. They were not within the human habitation rather the human habitation was within the *Srivanas*. These forests were managed locally. People depended on them for fuel, wood, medicines, fruits and many other day-to-day necessities. While people had access to these forests, they also had the duty to conserve, protect, replenish and enrich them. Enriching was done through planting of trees which provided medicines, timber, and fuel needed locally.

Srivanas had different types of trees, according to soil, climate and other environmental conditions. They were all, however, tuned to the social needs. *Tulasi* (basil) was found in abundance, for it was medicinal and at the same time it improved the fertility of the soil. These vanas abounded in fruit trees. The concept of social forestry was known to Indians even in that deep past, which we are now trying to import from the World Bank.

Five species of trees were a must in *Srivanas*. These were: (i) *banyan*, (ii) *peepal*, (iii) *ashoka*, (iv) *bela*, and (v) *harada*. *Banyan* is a self-generating plant. It does not die. In fact, it acquires the form of grove in due course of time. It is, therefore, associated with fertility. It is the abode of Lord Shiva and Devi Parvathi. It is shady, healthful and medicinal. *Peepal* (*ficus religiosa*) is a sacred tree. It is the only tree that gives oxygen for more than 20 hours a day. It is on this tree that Hindu ancestors reside. It is the incarnation of Lord Vishnu. Every part of peepal has medicinal value. Ashoka tree under which Sita Devi lived in Lanka, is a pain killer, as the name indicates. It is also a shady tree. *Bela* (aegle marmelos) bears fruits of great curative value it is useful in a large number of diseases. Its leaves are offered to Lord Shiva, that is why, wherever there is a Shiva temple, there invariably is a bela tree. The last of the five trees is *harad* (myro balem terminalia shebual). It is perhaps the most commonly used medicinal plant. It has great curative qualities and relevant for practically all diseases.

The name Panchavati is derived from these five trees. Every village had a Panchavati. The trees were not necessarily the same as listed above. They differed from region to region and sometimes from place to place and people to people. The other example is that of judicious use of water resources to improve the productivity of the ecosystem at the same time improving the sustainability of the environment.

Water Management

India has the distinction of developing unique irrigation systems as early as 3000 years ago which conserved and replenished surface and ground water and improved the living environment of the people. Those familiar with Deccan plateau, know that thousands of tanks doting this region around which villages are located provide an excellent example of how man can improve the environment. But for these artificial tanks, the plateau will remain barren and human existence precarious.

The annecutes, which our forefathers built across rivers in South India, point to what could be done to avoid the human tragedies of Narmada project, for example. Drinking Water wells dotted the villages and towns of India. The water of these wells was kept clean by underground replenishment to balance the water drawn for irrigation. There was no chance of contamination.

One can give several other hundreds of examples of water management, particularly wastershed planning and water harvesting. The choice of crops and trees depended on the climatic conditions particularly the rainfall. So even in Rajasthan desert, people were able to maintain forests. Our forefather knew the close relationship between trees, rainfall, water table and soil moisture. The cropping patterns were designed to conserve the water. Tanks were constructed to harvest water.

Concluding Remarks

So the message is clear. Humanity has no choice but to move towards a secure future and that it has to keep not of the imperatives as well as the potentials of sustainable environment. Countries like India have a great tradition of sustainable development. They also now have access to scientific knowledge and technological know-how developed in recent years. We need to blend the experimental knowledge of the past wit the scientific knowledge of the present not only to create a new knowledge but also, and more importantly, a new wisdom.

The task is difficult, as the analysis of economic, political and technological sustainability would show, but the Indian culture has a great resilience and Indian mind a great capacity of overcome hurdles. As the *Upanishads* say:

> "Get up, wake up and learn from the wise. As it is difficult to walk on the razor's edge, so is difficult to walk on the right path."

We have no start somewhere to come to the right path—the ecological path. Still it is not too late to make a beginning.

REFERENCES

Archer, A.A., *et al.*, 1987. *Man's Dependence on the Earth: The role of Geosciences in the Environment*, UNEP (Nairobi) and UNESCO (Paris).

Ciriacy, S.V., *et al.*. 1967. Naturla Resources: Quality and Quantity, Los Angeles, University of California Press.

Eyre, S.R., 1978. The Real Wealth of Nations, London, Edward Amold.

Goldsmith E., 1992. *The Way: An Ecological World View*, London, Rider.

Harrison J., 1927. *Themis: A Study Origion of Greek Religion*, Cambridge, Cambridge University Press.

Hira Lal, *Nature Cure*, (Manuscript in Press), 1993.

Mahatma Gandhi, *Hindi Swaraj*. Ahmedabad, Navajivan Press, several prints.

Mehta, R., 1982. *The Fullness of the Vid*, New Delhi, Motilal Baranasi Das.

Narang, S., 1993. *Atharva Vediya Prithivi Sukta*, New Delhi, Paridhi Praksan, 1993 (in Hindi).

NASA, 1969. *Resources and Man, San Francisco*, W.H. Freeman.

Pancavati, B., 1992. *Indian Approach to Environment*. New Delhi, Sri Vinayak Publications.

Patal, R., 1947. *Manand Temple in Ancient Jewish Myth and Ritual*, London, Thomas Nelson.

Swami Budhananda, 1988. *The Saving Challenge of Religion*, Madras, Ramakrishna Nath.

2

ENVIRONMENT (F)OR DEVELOPMENT
Exploring the Path for Sustainable Development

M.P. PARAMESWARAN

Introduction

TODAY there are many who pose the question "Environment or Development", being convinced that the two are mutually contradictory. Some of them take the side of environment and blame development is something undesirable. Others blame environmental consciences as it is considered to block development. Both of them are wrong. If we understand development in the proper perspective, if we do not mean it to be quick profit for a few, if we mean it as the sustained improvement in the living conditions of the majority, then environment is required for development. We are interested in environment for the sake of human beings.

Life, by its very definition, involves interaction with environment. All animals interact with their environment. But they do it in an unconscious way and to a limited extent. There exists a balance. Disruption of this balance leads to the extinction of the species. Human beings interact with environment in a conscious way. They go beyond the limits imposed by nature's balances, by consciously creating suitable environments and consciously transforming nature. Primitive human beings living in jungles, collecting fruits and roots and hunting animals transformed their environment only in a very limited way. Increased population, density, agriculture, industry—all these meant exponentially increasing rate of transformation of environment. The per capita consumption of energy in all its forms can be taken as a crude indicator of the rate of transformation of environment. When this rate goes beyond a limit the balances are upset and adverse effects shoot up. This can happen locally. There are indications of such local imbalances having occurred in the distant past. Gradually the extent of imbalance began to increase. Micro imbalances became macro in nature. Today things are becoming global. The favourable living conditions that man has created for himself are being threatened by the changes in the environment. To sustain these conditions we are forced to transform the environment even more drastically—thus leading to runaway or unstable conditions. This can result in the accelerated destruction of human life. How are we to prevent this? Even if runway conditions are not created, local conditions may so develop that a section of local society may find life intolerable. Chemical pollution of local air, water and soil, total loss of surface soil, total drying up of local water sources etc. are some of the examples leading to such local instabilities. One can quote ever so many concrete instances in our country and elsewhere.

The Three Environment Concept

We are interested in restoring stability, in maintaining stability and in preventing onset of instability. This requires a more detailed and deeper understanding about man and his environment. In order to achieve this, we propose to put forward "The Three Environment Concept". Human life is governed by three interconnected and inter related environments: The physical-biological environment, the socio-economic environment and the cultural environment. The existence of these three elements can be easily conceived.

The physical-biological environment can be, in turn classified into micro, macro and global.

Micro Environment

We are most sensitive to the changes in our immediate environment. The citizens of Bhopal and especially the residents of J.P. Nagar were the people who were immediately affected by the MIC leak from Union Carbide factory. Same is the case in Nagda or Harihar, Mavoor or Ennore. All these are examples of micro environmental impact. People's response to such impacts are quick and sharp.

Macro Environment

The deforestation of Himalayas or western ghats affected not only the local people, but the entire country. The hydrology is changing. Soil erosion causes silting up of stream beds and agricultural land. New water masses formed by artificial reservoirs change local temperature,— humidity regime and affect the vegetation pattern. Desertification tendencies have been reported. But all these are comparatively slow processes. People's response to these macro environmental impacts have been tardy.

Global Environment

The composition of the atmosphere is changing. Increased CO_2 content, it is alleged, has resulted in increased average global temperature. It is feared that ultimately this may lead to large scale melting of the polar ice cap, a rise in the mean sea level and inundation of vast tracts of low lying areas. The various chemicals that are released to the atmosphere make the air unbreathable; cause acid rains and even threaten the destruction of the ozone layer which protects us from the ultraviolet rays of the sun. The global radioactive fallout due to weapon testing or accidents in nuclear establishments, the spread of radioactive materials in the earth's hydrosphere, the pollution of the seas caused by oil tankers etc. are no longer to be neglected. Substantial debate on these global phenomena is taking place in the world press. But people, by and large, are unable to relate them to their own life.

Rachel Carsons "Silent Spring" is one of the most powerful micro-environmental studies. As early as 1876, Frederic Engles in the essay on "The Part Played by Labour in Transition from Ape to Man" had brought out the macro environmental impact of the destruction of the forests on the Alpine slopes. The global impact is a 20th century phenomenon, especially of the second half of the century.

The Three Mode of Existence

The mode of existence of human beings has been continuously changing. From gathering mode it changed to producing mode. The production changed forms. From primitive tribal form it has been transformed, through slave, feudal and capitalist to socialistic form. The corresponding socio economic forms can be basically classified into three categories: Primitive, Exploitative and Scientific.

Primitive

During the early days primitive human beings did not distinguish themselves from other elements of nature. The various customs, remnants of which have persisted even in the civilised society today in the name of *gothra* names and various taboos, indicate how primitive human beings related themselves to the nature around them. This micro environment itself was in stable equilibrium. Man's relationship with other men was cooperative. So was his relationship with nature. He was conscious of its depletion and destruction which he prevented through the various taboos.

Exploitative

The slave, feudal and capitalistic societies are exploitative in nature. So became man's relationship with nature. Temples and pyramids requiring large quantities of natural resources were built. Micro environments began to be disturbed in a large way. There were irreversible damages in many cases, especially due to lumbering and mining. Primitive human beings hunted for protection and food only. The feudal lords hunted for the sheer pleasure of killing. Tribal conflicts were part of the struggle for existence. Feudal wars were wars of annexation. With the development of competitive capitalistic economy demand on natural resource began to increase much more rapidly. Now onwards, to satisfy the same needs as in feudal time, one would require continuously increasing quantities of resource. Wars became wars for world hegemony. First macro environment and later even global environment became affected. Today man can irreversibly damage his life environment and lead to total destruction of all life, including himself, from the earth.

Scientific

Profit is the motive force propelling the capitalistic society. The Industrialists are not ready to purify their factory affluents whether at Mavoor or at Harihar or at Nagda. That will eat into their profit. Same is the case with the Hindustan Paper Corporation at Velloor, Kerala though in this case it is a public sector undertaking. A competitively organised society makes it impossible for the entrepreneur to care for the environment, to care for tomorrow. Further, to achieve a given social result in such a society many times more raw materials and efforts will be required than what is objectively warranted. If the relationship between man and man is unscientific, so will be the relationship between man and nature. Man is a relatively weak animal. Mankind survived through conscious co-operation. To overcome the present predicament too, mankind has to cooperate. Starwars cannot save mankind. No war can.

Following the Chernobyl nuclear accident (accidents can happen anywhere) the measures taken by the Soviet authorities to prevent further spreading of radiation, the massive environment management operation carried out by them, was unbelievable to the representative of USA, to the International Atomic Energy Agency. A profit dominated society cannot take enough measures to protect its environment. Even in socialist societies segmental interests—for example that of a given factory, interested in showing higher rates of profit—can go against larger interests. The two year long crusade that had to be mounted to stop the fouling of lake Ladoga, near Leningrad in Soviet Union is an excellent example. The arguments put forward by the factory management and the industries ministry were, word to word, similar to those put forth by the Birlas on the one had and the industries departments on the other hand. However, ultimately, they decided to close down the pulp factory and convert it into a furniture unit.

The Three Trend of Interaction

The situation in the underdeveloped or developing capitalist countries of the 'third world', like India is really hopeless. India is one of the most polluted countries of the world. Even her omnipurifying Ganga has become totally polluted: India has, also, to its credit the largest quantum of ecocrimes especially on the Himalayas and on the Western Ghats.

The mode of interaction of human beings with nature is governed by the manner in which they interact among themselves whether cooperatively or competitively. However, even in such societies this is not a static mode. Public opinion influences the modes dictated by the socio-economic environment and can lead to corrective intervention. The process of building up public opinion. The various conflicts involved in it, is termed by us as the Cultural Environment". It modulates the dictates of the socio-economic environment. Thus Ralph Nadar has become a legend in USA. So is the case with chipko Andolan in the Himalayas or KSSP in Kerala. All of them have fought against the current dictates of the prevailing socio-economic environment. In this cultural environment, too, there are three distinct trends: Nature Centered Profit Centered and Man Centered.

Nature Centre

This trend arose as a reaction to the blind and destructively exploitative attitude in the developed industrialized nations. They found that exponential growth has limitations. They found no other way to arrest the trend than to go back, to curb consumption. This gave birth, on the one hand to ideologues like Ivan Illich and Andre Garz and on the other hand to the very many shades of green Movements. Both were oblivious to the limitations imposed by the socio economic environment and the necessity to fight it. Andre Gorz argues that consumption is the culprit, that megatechnology is its source, and that 'socialism, being technofascism' will only accentuate it. There are ideologues like Susantha Conathilake (Aborted Discovery) and movements like PPST who take the simplistic view that western S&T with its exploitative relationship to nature, alone is to blame and that Indian and other Asian S&T are harmonious with nature and hence can offer the solution to the present predicament. They will oppose categorically nuclear energy, large hydel stations, large chemical industries etc. They advocate going back in history, back to nature. They do not perceive the impracticability and impossibility of their suggestions.

Profit Centred

This is the attitude of a socio-economic environment which demands total freedom for profit making and profit maximizing. Even those who oppose the existing socio-economic environment often fail to find the relationship between environmental destruction and super profits. Together they shout: large is beautiful, larger is more beautiful and largest is the most beautiful. The for them nuclear reactors any where, are as safe as match boxes. Fear of unseen radiation is superstition according to them. This is the profit centred cultural trend.

Man Centred

There is a third stream too. Today it is very tenous. They understand the compulsion of the socio economic environment, the limitations of the physical-biological environment and the urgency of change. They clearly see the costs and benefits involved in the various developmental activities. They know that the majority bears the costs and the minority reaps the benefits. They also understand that the solution is not giving back to nature, but a saner application of S&T for human affairs. The KSSP and many of the recent Peoples Science Movements in India belong to this group.

Fierce polemics have taken place and are going on in India among the groups representing these divergent trends. The Silent Valley debate heralded this, followed by the Tehri, the Narmada and many others. As far as pollution is concerned the issues are innumerable. Nuclear power and energy, in general, form yet another platform of polemics.

Concluding Remarks

It is the sum total of all these that constitute the cultural environment. We, who are active in this environment, who polemise on the man-nature interaction fully conscious of its repercussions on man-nature relationship would like to summarise our arguments as follows:

1. We are interested in nature and environment for the sake of human beings and human beings only.
2. Human society being stratified, what is advantageous for a few may be disadvantageous for the many. We take the side of the many.
3. What is advantageous today can be disadvantageous tomorrow and beyond. We do not want to sacrifices the future for the present.
4. We are also conscious of the fact that tomorrow can be reached only through today and solutions for todays problems are to be found.
5. We are confident that such solutions, which would not jeopardies tomorrow for today can be found.
6. History cannot go backward. Mankind cannot go back to primitive society. It has to go forward from competitive capitalism to cooperative socialism.
7. Exponentially increasing consumption is not compatible with finite resources on the earth.
8. However, presently known resources need not necessarily be the limiting factor. Science and Technology can make even sea water and ordinary mud versatile resources. Still they too are, theoretically, limited.
9. This does not mean that the underdeveloped third world countries are to stay where they are. They will have to, of necessity, increase their production several times so that the people can live in as human being.
10. Increase in production alone will not be sufficient to solve their problems. Control over production and distribution should be in the hands of the people. Right to make profit cannot be a fundamental right.
11. The present consumption pattern of the advanced capitalist countries is wasteful.

12. India is part of the capitalist world folk and cannot take totally independent decisions.
13. The impact of India's choices of development models on the global environment is negligible and cannot be the sole guideline for us.
14. However, different choices can produce different micro environmental impacts and even macro environmental impacts.

Similarly different technology choices can have different differential impacts on the various strata of the society. For example, a wagon factory on the banks of a river will have infinitely less local environmental impact than a pulp factory at the same place with the same investment or employment. The former is much more desirable. Similarly, a BHEL which can provide several times more employment and which use per worker less than one tenth of the scarce electricity is preferable to an Aluminium company. Construction of a Hill Highway can have a highly negative macro environmental impact. Expenditure of the same amount for afforestation, soil and water conservation measures and forest protection can have, on the other hand, highly salutary macro environmental impact.

15. Wherever such choices do exist, that which has the minimum negative impact has to be selected. Even in situations where at first look no such choice can be seen, closer scrutiny will reveal societally better choices. This might not be the best choice from the point of vested interest.
16. Wherever such choices do not exist, the benefits are to be compared with the total societal costs, taking into account the environmental impacts etc.
17. Laws of nature and laws of society cannot be changed. Having comprehended them man can use them to his advantage. There are social scientists and practitioners who show scant respect to the laws of nature; there are natural scientists who do not accept social laws. Both of them are bound to fail.
18. Lasting solutions can be obtained only with the change in the socio-economic environment so that blind exploitation of nature is no longer necessary. Struggle to preserve the quality of micro environment, threatened by the desire for excessive profits becomes, and should become, part of the struggle to change the socio-economic environment which makes destruction of physical-biological environment inevitable.

3

CAN SUSTAINABILITY BE MEASURED?

RICHARD A. CARPENTER

Measurements Relevant to Sustainability

SUSTAINABLE development depends largely on renewable natural resources of air, water, soil, sunlight, and communities of plants and animals. The utility, or potential for producing goods and services, of these natural systems is to be continued. The absolute amount of production is to be increased in order to maintain per capita flows since the human population will grow. These are the implications of sustainable development regardless of how arguments are settled about equities between or within generations, conversions among different kinds of capital, value judgements as to which goods and services are desired, or risk aversion as a function of economic status.

Nature, left alone, is self-organizing and is certainly sustainable in the sense that it has not discernible goals or purpose. Nature changes, responds, and evolves through a highly variable set of quasi-stable conditions over time. Human beings seek to impose some constancy and dependability of supply of needed products. Thus, sustainability now pertains to human management strategies, practices, or actions. Sustainability is relative, not absolute; a given management practice is more or less sustainable at a certain location over a certain time with a certain mix of products.

The most obvious measurement of sustainability is harvest over time. But the short term production of specific commodities may mask the loss of potential in the system to yield a broad range of goods and services over a long period of time. Hidden subsidies, management inputs, and borrowing from adjacent systems can maintain harvest levels.

To describe, in quantitative biophysical terms, the condition of a natural system that is responding to human intervention, is the first and essential step in operationalizing the concept of sustainable development.

At the present time, however, considerable uncertainty characterizes biophysical measurements relevant to sustainability (Carpenter, 1990). Once source of uncertainty is natural variation. Any "signal" that a change in condition is due to human action is often hidden in the "noise" of natural changes in the value measured. For example, no statistically significant change in the volume of water discharged by the Amazon river, or the amount of sediment delivered from the deforested Rondonia region, has yet been detected. A signal that deforestation has altered the hydrologic cycle or soil erosion in that basin is obscured by the high natural variability of rainfall, and El Nino events explain most of the occasional trends in the noise that are evident in the hundred of gaging stations (Richey, 1992). Other major contributors to uncertainty about sustainability are inadequate monitoring over both time and space, and sampling and analytical errors, especially under field conditions in developing countries.

Most important is the lack of fundamental understanding of cause-effect relationships and explanatory paradigms. For example, the atmosphere receives hydrocarbons and other combustible molecules from natural and anthropogenic sources. These materials are photochemically oxidized, in particular by the OH radical. Andreae (1992) notes the potentially unstable system in the remote atmosphere where methane and carbon monoxide are oxidized by OH which is consumed in reaction. A positive feedback occurs in that increasing amounts of CH_4 and CO consume more OH and are thus less rapidly oxidized and then build up to an even higher concentration. Methane is highly effective as a greenhouse gas and changing OH levels are a fundamental perturbation of atmospheric composition. Yet, andreae warns, "there are at present no reliable methods to measure OH on the time and space scales required to validate our current concepts and to detect a change in its atmospheric abundance and distribution".

Ecological research has not emphasized large area investigations, and that scale (regions such as the American midwest agricultural heartland, large marine ecosystems, or the boreal forests) is most important in monitoring and predicting sustainability of management practices in economic development. What studies have been made of ecoregions or biomes have revealed some of the sources of natural variation. Responses in nature to perturbations are non-linear, i.e., they involve lags, thresholds, and rapid transformations from one stable state to another. For example, despite considerable knowledge of atmospheric chemistry and years of monitoring the gradual build up of CFCs, the appearance of the Antarctic ozone hole was a surprise. Mixtures of positive and negative feedbacks, each poorly understood, or perhaps not even known, make predictions difficult. For example, only recently has forest management recognized the danger of building up accumulations of fuel on the ground through suppression of fires at the local level. Very long term effects sometimes become evident only after the original cause is no longer evident and so explanation is confounded (Magnuson, 1990).

The lament of the recent report of the Ecological Society of America (1991) is that:

> "Research programs exist to develop specific sustainable natural resources (e.g., sustainable forestry or sustainable agriculture). However, current research efforts are inadequate for dealing with sustainable systems that involve multiple resources, multiple ecosystems, and large spatial scales. Moreover, much of the current research focuses on commodity-based managed systems, with little attention paid to the sustainability of natural ecosystems whose goods and services currently lack a market value".
>
> "(Current) efforts are not presently united in a comprehensive research framework. Such a framework is needed because ecological processes link natural and managed populations to ecosystems and because common ecological principles underlie effective management strategies".

Expectations of Users of Biophysical Measurements

Economists, planners, and other development decision makers turn to natural scientists for data with which to run models, calculate return-on-investment, and devise polices to manage land, water and people. Their frustration in finding little, and highly uncertain, information is exemplified in this passage from a recent report. "Economic Policies for Sustainable Development" (Asian Development Bank, 1990):

> "In the preparation of this report no task proved more daunting than the assembly of reliable statistical indicators of recent trends and the current state of environment.
>
> "Aside from the most glaring cases where officials and researchers are aware of what is happening and can describe conditions in general terms most eloquently, the lack of quantitative environmental information comparable with the statistics available regarding

> economic parameters is a major obstacle to integrated economic-cum-environmental planning.
>
> "Statistical information on the environment is scarce, often inaccurate, seldom comparable from country to country, and rarely available in a time series covering a sufficient number of years to indicate trends in a reliable way. Thus, descriptions remain anecdotal and lack the hard edge of quantification which is necessary for analysis and policy formulation. . . .
>
> "Without question one of the most important findings of this study is that while reliable data on conventional economic parameters are plentiful, statistical information on the condition of the environment is scare and poorly organized".

Managers usually arrange for measurements that relate directly to their primary goal. Holling (1992) has studied a variety of managed ecosystems that failed or were unsustainable. He reports a sequence of events common to each of the systems studied. The management goal was always to reduce variability (for economic advantage) of some target parameter such as cattle grazing stocking density, fish population, or occurrence of fire or insects in forestry. In each case control was achieved as measured by the target variable. Natural capital increased and was stored. But slowly the total system changed to become more homogeneous, brittle, less resilient-an accident ready to happen. At the same time, management, having succeeded in its primary purpose, switched its concentration to improved efficiency in shifting cattle among paddocks, hatching fish, fighting fries and spraying insects. Measures of the status and trends of conditions in the whole ecosystem were too difficult or too expensive. The stabilized managed systems were attractive to investors and were further developed. These economic/social/ecological systems became increasingly vulnerable; to drought in pastures, inbreeding in fish hatcheries, contiguous forests that spread insect outbreaks and fires. Also, the other goods and services from the now simplified ecosystems declined. Ultimately, each system could not sustain production of the target commodity and the rigidified management and investment made reorganization an remediation more difficult. Measurements that might have guided adaptive management were not available.

Expectations by economists also include some metaphors that have so far not proved helpful. For instance, this conference is about "Investing in Natural Capital." Natural capital should somehow be analogous to financial capital and, therefore, if development can preserve natural capital and get along only on "natural interest," or if natural capital can be exchanged for an equal amount of technological or human capital, then a system is sustainable. It is, however, difficult to define natural capital, to even tell the difference between stocks and flows in some instances, to aggregate quite different forms of natural capital that also differ in size of scale boundaries. Components of natural capital vary greatly as to the time when degradation or depletion is evident, and in the time for recovery. For example, the lower atmosphere can be dirtied by a faulty damper in a flue but cleaned up with the next rainshower; leaching of toxic chemicals from a landfill can slowly contaminate a drinking water aquifer for decades before finally reaching an unacceptable level; salinization of soil through poor irrigation practices shows up slowly but may be rather quickly (if not cheaply) reversed; global warming has yet to be detected but carbon dioxide resides in the atmosphere for hundreds of years. It is not clear how such diverse stocks, degrees of degradation, and recovery periods can be usefully combined into one expression of natural capital that is roughly equivalent to money.

"Health" is another proposed analogy; i.e., a medical doctor checks a few "vital signs" and judges the health of an individual; so some similar minimum data set should succinctly describe the health of an ecosystem. But the human body is much more tightly integrated than any ecosystem. A reasonably functioning, productive, resilient ecosystem might actually be analogous to a person who reports to work with one arm, a recent heart transplant and bad case of the hives. The use of the term ecosystem health by Coastanza (1991), Schaeffer (1988), and others is more qualitative and vague, and perhaps more realistic than the temperature/blood pressure concept.

A final metaphor again illustrates the "economics envy" of some natural scientists-the ecological equivalent of Gross National Product. Proposals for a GNP-like single indicator of the sustainability status of a natural systems include Net Primary Productivity, as measured by remote sensing, and various Biodiversity Indices. Such integrating concepts could hide important qualitative differences among ecosystems that might show the same NPP or BI; and in fact, the GNP also is defective in this manner. For example, in forests there are "immense technical difficulties associated with measuring below ground productivity" (Franklin, 1992). And yet, the biomass of roots may have a high turnover rate. Ecology is unlikely to develop any simplified indicator of sustainability.

Measurement Problems in Managed Ecosystems

The following examples illustrate some of the practical difficulties in helping managers to get the most of what society demands from renewable resources without unacceptable damage to the production potential.

Agroecosystems

Food is so basic to human well being that sustainable agriculture is the oldest and best studied of all managed ecosystems. The year-to-year is the basic measurement, whether obtained by actual sampling or by interviewing farmers. In developing countries rice harvests are often never quantified but merely estimated and negotiated between farmers and buyers. False, low yields are reported when taxes or government shares are related to harvests. Experiment stations are often few and far between. Faeth, *et al.* (1991) note that "in the field, erosion-induced productivity changes are almost impossible to isolate and measure accurately".

Yield varies with weather and management which can obscure subtle changes in productivity. Much economic oriented information is available about agroecosystems but less is known about noncrop components such as soil organisms that ultimately affect sustainability.

The Consultative Group of International Agricultural Research Centres has added the wider aspects of sustainability to its goal of sustained production. Measurements are not available at the regional level to evaluate the extent to which agricultural practices degrade, maintain, or enhance the total ecosystem. Major research areas are the reversibility of degradation, thresholds of decline, and the biodiversity necessary for the future genetic base of agriculture (CGIAR, 1990).

Fisheries

As in agriculture the most important measure of sustainability is the level of catch and fisheries management organizations concentrate on estimating abundance of stock. There is less concern with measuring the level of depletion of a fish community below which its potential for recovery and production is lost. Measurements relating stock size and composition to environmental stresses such as pollution, or to loss of spawning habitat, are not well developed.

Even catch data are inadequate according to the International Centre for Living Aquatic Resources Management in the Philippines (MacKay,

1991). Underestimation occurs because as much as twice the marketable catch may be consumed or processed locally without being measured. In developing countries, data on the species composition of the catch are rarely available.

Sherman (1992) calls for standardized sampling by trawlers of the large marine ecosystems around the world. He notes that a consistent data base over many decades is necessary to judge whether over fishing is the cause of changes in species composition. Few areas have such records as yet, the North Sea and the Georges Bank being notable exceptions.

Temperate Forestry

The time between harvests in silviculture is usually many decades so there are few places where sustainability, even just of saw logs, has been demonstrated. For example, the block clear cutting practice in the US Pacific Northwest is now seen, after 40 years, as unsustainable. And for the future, American society is also demanding spotted owls and water for salmon from these forest lands. Franklin (1992) suggests that a reasonably assurance of sustainability of such a mix of forest products and services might require reducing the frequency of final harvest cut to 160 to 200 years from the current 80 to 100 years. This long rotation alternative management approach would reduce impacts on soil and water quality, and maintain greater biodiversity. This is indicative of (1) the substantial sacrifices of economic production in order to achieve sustainability, and also (2) the considerable uncertainty involved. While the lower return-on-investment would only be acceptable to government forest lands, even there the sustainability outcome would not be known for sure for perhaps 200 years.

Grazing Lands

Pickup and Morton (1992) describe a method using remote sensing in arid lands to separate grazing-induced changes in forage biomass from rainfall-driven variations. "Grazing gradients" are patterns developed because sheep and cattle graze out only so far from water sources, and the areas where the vegetation is consumed can be detached. If the grazing gradient does not disappear with the recovery of vegetation after large infrequent rainfalls, the land has been more or less permanently damaged. If recovery does occur once, it does not mean that site is permanently sustainable. Large, unpredictable fluctuations in rainfall, and explosions of weeds and mammalian pests may yet ruin that pasture. They conclude that for Australia, "Meshing of wealth-generating uses of land with ecological sustainability remains as unachieved goal".

Lusigi (1992) points out that African tropical range lands are not actually a managed production system but are responding to population pressures and political disruptions by shifting from a nomadic (plausibly sustainable) mode to a sedentary situation which is obviously rapidly degrading entire regions.

Risser (1992) suggests measures of range condition including biodiversity of plant species, peak standing crop, nitrogen content of foliage, and soil organic matter. These data would be difficult to obtain in developing countries. Opportunities for Improving Biophysical Measurements.

The scale for managing (and therefore for measuring) sustainability is at the landscape or regional level of thousands to millions of hectares. Local harvest-related data cannot be aggregated to give the needed information. Land use changes appear to be a fundamental cause of reduced ecosystem function and reduced sustainability of production of the goods and services desired. But ecosystems are naturally changing and adapting so that measures must discriminate effects of human activity against this background of change.

Over half the earth's surface has been transformed from its natural state by agriculture, forestry, urbanization, desertification, and other interventions. The altered landscape affects adjacent aquatic systems. One third of the loss of species is ascribed to land use change.

One new measurement approach is to combine growing capabilities in remote sensing, geographic information systems, and landscape ecology. The product is a set of data depicting landuse change that is practical, sensitive over time, and interpretable in terms of sustainability (O'Neill, 1992). Map-lie outputs readily communicate to decision makers about habitat coverage, ecotones, patch configuration, economic activity, water quality and vegetation.

Supporting these measurements of landscape composition and pattern should be a selected data set for the particular type of managed ecosystems under consideration. Site specific measurements (although individually unreliable for determining regional sustainability) can be gathered into status and trend reports. Integrating organisms and keystone species can be monitored; e.g., lake trout in Lake Superior. Measurements now plausible that could be standardized for comparison throughout the world include: (1) yield of primary goods and services (harvest);

(2) hydrologic cycle; (3) water quality; (4) soil "report card"; (5) tropospheric acid gases and ozone; and (6) keystone and pest species populations. The research and monitoring agendas of management agencies and scientific organizations are beginning to focus on the broader conditions of ecosystems that relate to their sustainable utility.

Summary: Can Sustainability be Measured?

The answer to the question posed by the title of this chapter is: The capability to detect unsustainability far exceeds the capability to confirm or predict sustainability with confidence. This sobering conclusion is based on an extensive survey of recent literature (see Carpenter, 1992, for excerpts), field work in Southeast Asia by various projects under the auspices of the East-West Centre (Rerkasem and Rambo 1988; Carpenter and Harper 1989), and the recent conference on Definition and Measurement of Sustainability; the Biophysical Foundations, held at the World Bank in Washington on June 22-25, 1992 (UNU, 1992).

Sustainable development is being formulated, elaborated, interpreted and debated in the abstract, but if it is ever to be implemented there must be a basis of relevant, statistically reliable, scientific measurements. Such quantifications will (1) describe the conditions of natural systems under various management practices; (2) monitor trends in the conditions and responses to natural and human-induced stresses; and (3) predict the outcome of proposed alternative management practices. Only with a foundation of useful biophysical measurements can the legant models of ecological economics proceed to inform and influence policy and decision-making.

Science is not completely incapable of measuring sustainability, and gross and obvious unsustainable practices need no measurements for their evaluation. Many subtle evidences of unsustainability also can be detected. Measurement inadequacies and uncertainties about ecosystem conditions are most important, however, when demands for goods and services from renewable natural resources are large and urgent, and the question is the maximum sustainable yield. To err in overstressing the natural production system will likely cause its partial, at least temporary, collapse. To err in foregoing too large a portion of yield in order to protect the basic resource will likely deprive some poor people of food and other needed natural products.

Measurement difficulties arise from (1) Ignorance of descriptors of the condition of ecosystems at the regional scale; (2) Non-linearity of ecosystem response to stress; (3) Natural variability (signal-to-noise ratio); (4) Sampling and analytical errors under field conditions; (5) Inadequate monitoring over space and time; (6) Deliberate falsification of data for socio/cultural or political reasons; and (7) Inadequate research on the linkage of natural and managed ecosystems.

Opportunities to overcome these difficulties through improved monitoring and understanding of the structure and function of ecosystems are at hand. International development assistance agencies should recognize the current state-of-the-science of biophysical measurement and provide appropriate support to improve the situation. Such funding would appear to be highly cost-effective because better measurements will permit the realization of a greater productive potential from natural resources and, perhaps, sustainable development.

REFERENCES

Asian Development Bank, 1990. Economic Policies for Sustainable Development, page(s) i-xx, 1-253. Asian Development Bank, Manila, Philippines.

Carpenter, R.A. and Harper, D.E., 1989. Towards a science of sustainable upland management in developing countries. *Environmental Management* 13: 43-45.

Carpenter, Richard A., 1990. Biophysical Measurement of Sustainable Development. *The Environmental Professional*, Vol. 12, pp. 356-9.

Consultative Group on International Agricultural Research (CGIAR), 1990. Report of the Committee on Sustainable Agriculture. Consultative Group Meeting, May 21-25, 1990. The Hague, The Netherlands.

Constanza, R., (Editor), 1991. *Ecological Economics: The Science and Management of Sustainability*, Columbia University Press, New York.

Ecological Society of America, 1991. The Sustainable Biosphere Initiative: An Ecological Research Agenda. *Ecology* 72: 371-412.

Faeth, P., Repetto, R., Kroll, K. Dai Q. and Helmers, G., 1991. *Paying the Farm Bill: U.S. Agricultural Policy and the Transition to Sustainable Agriculture*, page(s) i-x, 1-70. World Resources Institute, Washington D.C.

MacKay, Kenneth Tod, 1991. (International Center for Living Aquatic Resources Management). Global Warming, Fisheries and Policy for Sustainable Development. Presented at the International Conference on Global Warming and Sustainable Development, Bangkok, June, 10-12, 1991.

Magnuson, J.J, 1990. Long-term ecological research and the invisible present. Uncovering the processes hidden because they occur slowly or because effects lag years behind causes. *BioScience*, 40:495-501.

Rerkasem, K. and Rambo, A.T., 1988. *Agroecosystem Research for Rural Development*, page(s) i-vi, 1-287. Multiple Cropping Centre, Faculty of agriculture Chiang Mai University and SUAN, Chiang Mai, Thailand.

Schaeffer, D.J., Herricks, E. and Kerster, H., 1988. Ecosystem Health I, Measuring ecosystem health. *Environmental Management* 12(4), pp. 445-55.

UNU, 1992. Papers presented at the United Nations University. International Conference on the Definition and Measurement of Sustainability: the Biophysical Foundations. The World Bank, Washington D.C., June 22-25, 1992.

4

POLITICS AND ECONOMICS OF RIO CONFERENCE

T.M. THOMAS ISSAC

THE Rio Conference signifies a global recognition of the worsening of poverty, ecological crisis and the imperative need for the integration of environment and development concerns so as to render development a sustainable process. At the same time, the Rio Conference was a mirror of the contradictions which characterise our epoch and conflicts arising therein. The North-South divide confounded every issue. There were conflicting interpretations of the causes, nature and remedies for the global environmental crisis. Consequently there is a danger of one sided assessment of perhaps one of the most important events in our times. On the one hand, it has been maligned as the supreme example of 'green imperialism' and, on the other, hailed as the harbinger of a new ecological era. We intend to bring out the complexity of the process at work through a textual analysis of the basic document of the Conference, particularly, the Agenda 21.

This voluminous document whose draft extend over 600 pages has over 40 chapters presented in four parts. Part I contains the Preamble (Ch. 1) and discusses 'the social and economic' dimensions of the global environmental crisis (Chs. 2-8) such as poverty, consumption patterns, population, health, human settlement patterns and policy aspects of integration of environment and development in decision-making. The major problems of 'conservation and management of resources' in various sectors such as atmosphere (Ch. 9), land resources, forests and other fragile ecosystems (Chs. 10-13) and oceans and freshwater are taken up in Part II. Some of the key issues arising from the utilisation of the above resources like sustainable agriculture, biodiversity and biotechnology (Chs. 14-16) and waste management (Chs. 19-22) are also discussed. Part III considers 'strengthening the role of the major groups such as women, youth, indigenous people, trade unions, farmers, business community (sic.) and scientists (Chs. 24-26, 29-32) in implementing the programme on environment and development. Role of local authorities and non governmental organisations also receives separate treatment (Chs. 27-28). Finally in Part IV 'the means of implementation' namely scientific and technological (Chs. 34-35), educational (Ch. 36), legal and institutional (Chs. 37-40) are taken up. Means of 'financial resources and mechanisms' (Ch. 33) could not be finalised and was excluded from the draft text. Each chapter identifies a number of programme areas which are discussed in terms of basis of action, objectives, activities and means of implementation. Portions in the text which required further negotiation at the Conference were placed in square brackets. Our discussion in this chapter is entirely based on the draft text as the final version of the document as adopted by the Conference has not been available to us.

We shall not take up the chapters in the order of the text for critical evaluation but organise our discussion around certain major themes namely market and the environment (Section I), poverty and overconsumption (Section II), biodiversity and biotechnology (Section III), pollution and climate change (Section IV), and environment and development (Section V). 'The Convention on Biological Diversity' and 'Convention on Climate Change' are taken up in Sections III and IV respectively. The other two important documents of the Conference 'Forest Principles' and 'Rio Declaration on Environment and Development' are only touched upon. In order that the thread of argument may not be lost in the labyrinth of presentation, three major points that underlie our analysis may be presented at the outset itself.

The first point to be made in any assessment is that the Rio Conference was a statement of global consensus of the reality of the environmental crisis. To some, perhaps, the above statement may appear trivial. But it would cease to be so if one reflects on the less than half a century history of the environment movement. Environmental problems, till not so far back, were the issues of only a small number of concerned scientists and fledgling environmental groups. It is not surprising that the capitalist system that revolves around private profit motive did not consider social costs and benefits in its calculus. Equally surprising was the inadequate sensitiveness of the anti capitalist class movements to the environmental concerns. At a philosophical plane, the fault lay in a rather one sided and undialectical understanding of the man nature interaction. At the level of practice, it reflected in a reluctance to look beyond immediate economic interests and gains. The astoundingly high pollution levels in some of the East European countries and the litany of monumental environmental follies in the erstwhile Soviet Union, from deep ploughing the 'virgin lands' to drying up of Aral lake and reckless attempts to divert Siberian rivers are today well known. Traditions of Lysenkoism die hard. It is in this sense the Rio Conference marks a turning point . The conference was bitterly divided on many issues. But there was universal acceptance of the fact that the environmental crisis is one of the grave dangers confronting humankind today.

The above global consensus was emerging in an international context where the world power balance had shifted definitely in favour of imperialism with the collapse of East European Socialist Bloc. This has left a deep imprint on the Rio Conference and its documents. There was a concerted effort by America to convert Agenda 21 into a eco counter part of the Dunkel Draft Text in trade. The developed countries and their transnational corporations refused to pay for the pollution they created. Their attempt was to shift the entire burden of adjustment necessitated by the ecological crisis onto the shoulders of the Third World. These attempts were contested by the Third World countries and caught widespread media attention. However, torn between conflicting interests the Rio documents were rendered largely toothless and thus a litany of well intentioned but often contradictory statements. Agenda 21 is significant for not only what it has said but also for what it has left unsaid or stated in square brackets. How else could it be otherwise when the TNCs themselves had assumed the role of patrons of the Conference? Even the apolitical environmental groups were forced to come out in the open against the American maneuvers. Rio has put a final nail on illusions that ecological problems are above class relations and class struggle. Any genuine environment movement in today's context would necessarily have to be anti imperialist. This is the second point that we wish to highlight.

Does the refusal of the Third World countries to accept unilaterally imposed green conditionalities imply that the environmental concerns are irrelevant in the Third World? There could not be a more a naive or apolitical understanding. The traditional dichotomy between development and ecology is false. The burden of environmentally degrading economic growth falls on the poor and the future. It is not the polluter but the polluted who pays—not only internationally but also within nations. Therefore any genuine people's movement can ill afford to make a fetish of economic growth ignoring national class and ecological realities of the growth process. It is paramount that there has to be an acceleration of economic growth in Third World. But this economic growth should be equitable and sustainable. Many of the principles of sustainable development enunciated in Agenda 21 can contribute to the reorientation of the growth process.

Section I: Market and the Environment

The contemporary environmental crisis is the supreme example of the failure of the market. Free market cannot account for social costs and benefits. Phenomenal expansion of such externalities is characteristic of the growth of modern productive forces and an indication of the latter's growing socialisation. The private appropriation is constantly at loggerhead with the social nature of production. This was a powerful argument for social intervention and control of modern economy. So far such calls have been largely ignored but the rising tide of environmental consciousness will not permit such laissez faire attitudes any more.

It is in the background of the above that concerted efforts are being made to evolve an environmental economics that is 'market friendly'. The chief objective of the effort is to make prices reflect social costs and benefits. There are three aspects to it. First is the elimination of subsidies for resource use. Subsidies are criticised to be wasteful and inefficient. The most frequently quoted examples of the World Bank are energy and agricultural inputs in the Third World. Second is the polluter pays principle, adopted by OECD in 1972, where the cost of measures to reduce pollution incurred by public authorities is collected from the polluter through tax or special fee. But the preferred mode by the World Bank is the pricing of additional pollution or additional resource use through economic incentives such as tradeable permits for pollution emissions. Third is creation of markets in open access resources through clarification and enforcement of property rights. Only as a last resort should regulatory approaches such as setting standards, fixing quotas, zoning areas, etc. be implemented. Failure of central planning to protect environment in Easter Europe is also a standard reference.

There is no doubt that price and fiscal measures can play a supplementary role in environmental regulatory regime. But what needs to be contested is the transubstantiation of the World Bank argument somewhere down the line from the need for market incentives and disincentives to one where international free trade regime as modelled by GATT as the guarantee for environment protection. In the latest

World Development Report on environment, World Bank unabashedly claims:

> "Fortunately, many policies that are good for efficiency are also good for the environment . . . World Development Report 1991 described a set of 'market friendly' policies for development. These included investing in people through education, health nutrition, and family planning; creating the right climate for enterprise by ensuring competitive markets, removing market rigidities, clarifying legal structures, and providing infrastructure; fostering integration with the global economy through promotion of open trade and capital flows; and ensuring macroeconomic stability. All these policies can enable better environmental management." [World Development Report 1992, 1992:10]

The World Development Report, 1992 was itself the outcome of one of the many frantic attempts of the USA and the World Bank to decisively intervene in the global environmental negotiations and subvert them from reaching their logical conclusions which would have been inimical to the new world order being stringed together under GATT negotiations. These efforts have been successful in incorporating the global free trade regime as the basic framework for Agenda 21. In fact, the Agenda 21 denotes a major departure from the traditions of its preceding international commissions such as Willy Brandt Commission (1980) ("North-South: A Programme for Survival") or Brundtland Commission (1987) ("Our Common Future") not to mention the various reports of the South Commission in explicitly accepting an international free trade GATT order, highly loaded against the third world, as the immutable world order of development. Chapter 2 of the Agenda seeks to promote "sustainable development through trade liberalisation" [Agenda 21, 2.3(a)].

Such hopes are neither borne out by theory nor experience. The new international order is inimical to both economic growth and environment of the Third World. It is significant that despite its obeisance to GATT negotiations, Agenda 21 also presents severe indictment of the same on the basis of examples from real historical experience of the Third World. A selective reading of the text itself may present a comprehensive set of criticisms of the new GATT order:

> "The reactivation and acceleration of development requires both a dynamic and supportive international economic environment and determined policies at the national level. . . .The development process

will not gather momentum if the global economy lacks dynamism and stability and is beset with uncertainties. Neither will it gather momentum if the developing countries are weighted down by external indebtedness, if development finance is inadequate, if barriers restrict access to markets and if commodity prices and the terms of trade of developing countries remain depressed. The record of the 1980s was essentially negative on each of these counts and needs to be reversed." [Agenda 21, 2.2]

"An important feature of the world commodity economy in the 1980s was the prevalence of very low and declining real prices for most commodities in international markets and a resulting substantial contraction in commodity export earnings for many producing countries. The ability of those countries to mobilize, through international trade, the resources needed to finance investments required for sustainable development may be impaired by this development and by tariff and non-tariff impediments, including tariff escalation, limiting their access to export markets. The removal of existing distortions in international trade is essential." [Agenda 21, 2.7]

"Many developing countries have experienced a decade-long situation of negative net transfer of financial resources, during which their financial receipts were exceeded by payments they had to make, in particular for debt-servicing. As a result, domestically mobilized resources had to be transferred abroad instead of being invested locally in order to promote sustainable economic development." [Agenda 21, 2.23]

"For many developing countries, the reactivation of development will not take place without an early and durable solution to the problems of external indebtedness, taking into account the fact that, for many developing countries, external debt burdens are a significant problem. The burden of debt-service payments on those countries has imposed severe constraints on their ability to accelerate growth and eradicate poverty and has led to a contraction in imports, investment and consumption." [Agenda 21, 2.24]

And finally on structural adjustment programmes:

"While such programmes are necessary for improving the balance in fiscal budgets and balance-of-payments accounts, in some cases they have resulted in adverse social and environmental effects, such as cuts in allocations for health care, education and environmental protection. It is important to ensure that structural adjustment programmes do not have negative impacts on the environment and social development so that such programmes can be more in line with the objectives of sustainable development." [Agenda 21, 2.35]

The criticism against structural adjustment programmes is soft indeed. It is not instance of "some cases" but a near universal experience that the structural adjustment programmes seek to restore macroeconomic balance at the expense of economic growth and natural resource stock. The continents of Latin America and Africa experienced negative growth rates during the entire 1980s. Unrestricted free trade has historically been the major channel for siphoning of Third World's resources. The structural adjustment programmes add new compulsions to the above. Unrestricted free trade can result in mono crop specialisation which would be a grave threat to biodiversity. But any criticism of adverse impact of trade on environment were carefully edited out of the text of Agenda 21 at the insistence of USA.

It is a supreme irony that the World Bank and USA do not consider it necessary to apply the much touted market friendly principles for environmental protection to the international free trade regime. The polluter pay principle seems to have no validity in the international context. Nor does the price mechanism to conserve resources. Will not a significant increase in the price of Third World commodity exports conserve the scarce resources? Will not private property principle have application to the genetic resources in the Third World? The answers to each of the above questions will bring out the American double standards.

Yet another major weakness of the Agenda 21 document is its silence on the role of transnational corporations in ecological degradation. There is a clear trend to soft pedal the need for international monitoring and regulation of TNCs and play up the willingness of these corporations for self regulation. The document "Transnational Corporations and Sustainable Development", prepared by United Nations Centre for Transnational Corporations (UNCTC), which presented a detailed code of regulations that TNCs are obliged to follow was shelved. UNCTC has itself been shut down as a part of ongoing restructuring of UNO. In its place the Business Council for Sustainable Development consisting of leading corporations was promoted. This Council succeeded in muting criticisms against TNCs and in inserting a section in Agenda 21 "strengthening the role of business and industry" for environmental protection (Agenda 21, Ch. 13). It is symptomatic of our times that even the greatest polluters are forced to adorn the garb of environment protectors.

Section 2: Poverty and Overconsumption

The contemporary global ecological crisis is the direct outcome of colonial and neocolonial exploitation of the Third World and the unsustainable pattern of production and consumption in the developed capitalist countries. The developed countries have little over one billion people while the developing countries account for nearly four and a half billion people. But the industrialised countries consume 70 per cent of the world commercial energy, 75 per cent of the metals, 85 per cent of the wood and 60 per cent of the food. North consumes five-sixths of the world's resources that are used. Thus the average consumption in the North is around 20 times the average consumption in the South. The difference has been widening over time. In 1889 the real income per person in Europe compared with a person in China or India was only 2 is to 1. But in 1991 it has increased to 70:1 [New Internationalist, September, 1992]. This historical trend is continuing: "The absolute differences in per capita income between the top 20 per cent and bottom 20 per cent of world population, expressed in 1989 US dollars, increased between 1960 and 1989 from $1,864 to $15,149. These disparities are reflected in real consumption levels." [UNDP, Human Development Report, 1992]

Today the poorest 20 per cent of the world have only 1.4 per cent of global income while the richest 20 per cent account for 82.7 per cent of the income. It is this international inequitable path of development that is unsustainable. Third World's demographic problem has often been used as an ideological diversion from the stark reality presented by the above unassailable facts. The world population is expected to more than double before it stabilises. Most of the increase is expected to occur in the Third World countries. Such an increase would put severe stress on the finite resources of the world. Therefore to many of the neo-Malthusian breed, population growth is the basic cause of the present crisis and the early stabilisation of population is the main to the resolution of the crisis.

In this paradigm, "the poor are both victims and agents of environmental damage" [World Development Report, 1992:7, 30]. The ideological intent of these victim blaming theories is self evident. More resources would certainly be required not only to meet the basic needs of the growing population but also to improve the living standards of the poor. But the reason why they are being met today in a environmentally destructive manner is to be sought in the logic of the world capitalist system.

Draft of Agenda 21 reveals an awareness of the routine ideological trap, as the following two quotes amply prove (even though they are placed in square brackets):

> "While poverty largely results in certain kinds of environmental stress, one of the most serious problems now facing the planet is that associated with historical patterns of unsustainable consumption and production, leading to environmental degradation, aggravation of poverty and imbalances in the development of countries." [Agenda 21, 4.3]
>
> "Special attention should be paid to the demand for natural resources generated by unsustainable consumption and to the efficient use of those resources consistent with the goal of minimizing depletion and reducing pollution. Although consumption patterns are very high in certain parts of the world, the basic consumer needs of a large section of humanity are not being met. This inequitable distribution of income and wealth results in excessive demands and unsustainable lifestyles among the richer segments, which place immense stress on the environment. The poorer segments, meanwhile, are unable to meet food, health care, shelter and educational needs. Changing consumption patterns will require a multipronged strategy focusing on demand, meeting the basic needs of the poor, and reducing wastage and the use of finite resources in the production process." [Agenda 21, 4.5]

What is forgotten by population fetishists is that the Third World population problem is not merely a demographic phenomena but also structurally determined one. Though no one would debate the need for consciously integrating the demographic variable into development planning, one sided and unilateral interventionist strategies forgetting the socio-economic determinants of demographic transition would only be counter productive. Agenda 21 has a more holistic view of the demographic question and seeks to integrate with other human resource sectoral policies such as health, education and other basic needs. It seeks to formulate "integrated national policies for environment and development, taking into account demographic trends and factors." (Agenda 21, Ch. 5)

But there was no agenda to tackle the historic trends of unsustainable exponential growth consumption pattern of the Northern countries. Without such a reduction and redistribution of world income to ensure improvement in the standard of living in the South and bridging of the

North-South gap the world cannot move into a sustainable growth trajectory however catastrophic the present trends are. But as the president Bush declared: American life style was not for bargaining! This failure constitute the weakest link in the Agenda 21.

Section 3: Biodiversity and Biotechnology

Loss of biodiversity—variety among living organisms and the ecological conditions they inhabit—is one of the grave global threats arising from disruption of areas of high diversity such as rain forests, temperate rain forests, coral reefs, freshwater lakes and homogenization caused by modern agriculture. "The genes, species, ecosystems, and human knowledge that are being lost represent a living library of options for adapting to local and global change. Not only are species being sentenced to extinction, but the reduction and fragmentation of habitats prevents the evolution of new species." [World Resources, 1992-93, p. 127] It is estimated that the continued loss of species at the present rate could zoom up to 15 per cent of the earth's species over the next 25 years with severe potential losses of advances in medicine and agriculture. The homogenization of genetic variety in the domesticated species which is also taking place at an astounding rate has severe implications for global food security.

The movement for preservation of biodiversity has received further impetus with the advent of biotechnology. Biodiversity is the basic raw material for biotechnology, the engine of future industrial revolution. Biodiversity is often described as the "green oil" of the future. A global struggle for its control is underway. The habitats rich in biodiversity are in the South while the advances in biotechnology are the preserve of the North. The US corporations are the technology leaders. So far they had free access to the genetic resources of the South either through direct forays into the habitats or the gene banks sponsored by the World Bank. The Third World countries were neither compensated for the genetic resources nor allowed to share in the profits generated by them. The biotechnological innovations remained in the close control of TNCs. This basic inequity in the North South relations was an important point of contention at the Rio Conference.

There was unanimity in the need to preserve biodiversity. But the game plan of USA involved: (a) guarantee of continued free access to the gene resources by ensuring 'global management' of key habitats to be declared as global heritages; (b) assurance of control of biotechnologies by TNCs through universalising American patents system; and (c) effective regulation of funds that would have to be allotted for the preservation of ecosystems in the Third World. US strategy was stated in the clearest terms by the Heritage Foundation who emphasised "the enormous stake in maintaining the integrity of intellectual property" and the need for opposing any UNCED proposals regarding transfer of "environmentally beneficial technologies." It also warned the US negotiators of any action by the UNCED that would interfere with the intellectual property agreement that may emerge from the Uruguay Round. Further instead of any additional funding for environmental protection, the US should insist on "free-market reforms." [Vandana Shiva, "Hijacking the Earth", *Frontline*, July 17, 1992]

The interest of Third World countries were diametrically opposed to the USA. What they required were: (a) an assertion of national sovereignty over genetic resources; (b) access to biotechnologies at favourable terms; and (c) untied funds for preservation of the habitats.

To what extent does the Rio documents reflect and accommodate the above conflicting interests is the issue that we have to analyse. To start with, the Third World countries succeeded in neutralising the concept propounded by USA that the gene resources were a global heritage. Agenda 21 acceded to the demand of the Third World countries for recognition of their national sovereignty over their genetic resources:

> "At the same time, it is particularly important in this context to stress that States have the sovereign right to exploit their own biological resources pursuant to their environmental policies, as well as the responsibility to conserve their biodiversity and use their biological resources sustainably, and to ensure that activities within their jurisdiction or control do not cause damage to the biological diversity of other States or of areas beyond the limits of national jurisdiction." [Agenda 21, 15.3, See also Convention on Biological Diversity Article 3]

What is the implication of the national sovereignty over biological resources? It means recognition of the "sovereign rights of states over their natural resources, the authority to determine access to genetic resources rests with national governments and is subject to national legislation". (Convention on Biological Diversity, Article 15.1)

Vandana Shiva has pointed out three basic flaws to the concept of Third world sovereignty over the biological resources as incorporated in the Biodiversity Convention. Firstly, the Convention is silent regarding the status of gene samples in the already existing gene banks that reportedly cover 90 per cent of the known genetic resources of most important agricultural crops. Most of these collections are under the control of developed countries or World Bank. Secondly, it does not recognise "the sovereign right of local communities which have conserved and maintained biodiversity. . . ." And, thirdly 'countries of origin' of genetic resources in Article 15 that are eligible for certain special rights in receiving some of the benefits of the use or development of those resources by other countries (as per Article 16 and Article 19) have been defined in Article 2 in such manner as to give rise to claims by countries housing botanical gardens and laboratories in which resources are located or developed. It thus strengthens the bargaining position of developed countries as they could also claim to countries to origin by nature of their possession of gene banks or botanical gardens. [Vandana Shiva, 'The Diversity Convention and Patents. Did India Vanes' *Earth Summit Briefings*, Third World Network]

The first criticism is certainly a major flaw of the Biodiversity Convention. The exclusion of the existing gene bank resources from the Convention would be a further fillip to the tendencies already evident for the privatisation of these public resources collected by international agencies on the understanding that they would be easily or freely available for public use. World Bank controlled Consultative Group on International Agricultural Research (CGIAR) has already announced its intention to patent some of their genetic materials and collaborate with MNCs in their utilisation. The technological gains from these resources would flow to the developed countries without any compensation to the developing countries who conserved or developed and provided these materials to the international agencies.

We are in agreement regarding the need to recognise the rights of local communities or indigenous population as well as the need to compensate for their knowledge and contribution to conservation. But "sovereign rights" can be vested only in the national governments. Not only is the alternative interpretations brought with serious dangers of separatism but also with problems of operationalisation.

The 'country of origin of resources' should have been defined unequivocally. Now it is defined as "the country which possesses those genetic resources in *in-situ* conditions". And, '*in-situ* conditions' as "conditions where genetic resources exist within ecosystems and natural habitats and in the case of domesticated species or cultivated species, in the surroundings where they have developed their destructive proportion".

Finally, 'eco-system' is defined as a "dynamic complex of plant animal and micro organism communities and their non-living environment interacting as a functional unit". (Article 2) Now, whether the absence of the term "naturally existed" or "naturally developed" or the replacement of the term "ecological unit" by "functional unit" would imply that laboratories and botanical gardens also could seek coverage in "country of origin" is debatable.

Even though severely contested by America (and for that reason square bracketed), Agenda 21 provided for transfer of biotechnologies to the Third World under fair and favourable terms:

> "Take appropriate measure for the [fair and equitable] sharing of benefits derived from research and development and use of biological and genetic resources,including biotechnology, between the sources of those resources and those who use them;" [15.4(e)]
>
> "[Develop measure and arrangements to implement the rights of countries of origin—areas of origin and/or natural diversification—of genetic resources, particularly developing countries, to benefit from the biotechnological development and the commercial utilization of products derived from such resources]." [15.4(j)]
>
> "[Provide for and/or facilitate the transfer of and cooperation on technologies relevant to the conservation of biological diversity and the sustainable use of biological resources or technologies that make use of genetic resources and cause no significant damage to the environment, recognizing that technology includes bio-technology]." [15.7(d)]

In the final version article 15.7(d) was severely watered down and article 15.4(g) made conditional to 'without prejudice to the relevant provisions of Conventions on Biological Diversity'. The treatment of the issue in the Biodiversity Convention is unsatisfactory from the point of view of the third world. The Convention provides for access to and transfer of biotechnology "under fair and most favourable terms, including on concessional and preferential terms where mutually agreed, and where necessary in accordance with the financial mechanism established by Articles 20 and 21". (Article 16.2) However it also states that "in the case of technology subject to patents and other intellectual property rights, such access and transfer shall be provided in terms which recognise and are consistent with adequate and effective protection of intellectual property rights (IPRs). (*Ibid.*)

The reference to IPRs in the background of the then ongoing Uruguay Round talks too certainly prejudicial to the Third World countries. But to conclude from it that biodiversity treaty had prejudged the Uruguay Round on IPRs is unwarranted for the following reasons:

(a) As far as our reading of the available text goes, the reference to IPRs in the Convention is with respect to technologies that have to be transferred to the developing countries. Such transfer would be subject to the patent laws of developed countries. Transfer of technology can only be on mutually agreed terms. Therefore, what the above amendment implies is that developed countries have preserved their discretion to permit concessional access to technology. However, the negotiations in the Uruguay Round with respect to IPRs was of a different order altogether. It was concerned with the patent laws within the developing countries that had been framed with a view to promote indigenous technological growth. Americans want universalisation of patent laws that are primarily intended to protect the interests of transnational corporations. Absence of reference to IPRs in the biodiversity convention even with reference to technology transfer would have strengthened the hands of Third World negotiators in the Uruguay Round. But its reference had also not tied their hands in anyway in defending our patent laws against the American onslaught in the Uruguay Round.

(b) Even with respect to technology protected by patents and other intellectual property rights clause 16.3 of Bio-Diversity Convention enjoins the contracting party to take legislative, administrative or policy measures for technology transfer to the developing countries "on mutually agreed terms" where necessary through financial mechanisms established by the Convention.

(c) Further it has been argued by Ashish Kothari that the Biodiversity Convention could even be used as a potential aid in the patent negotiations because yet another clause in the Biodiversity Convention requires that the IPRs do not run counter to the objectives of the Convention. ". . . recognising that patents and other intellectual property rights may have on the implication of this Convention, shall cooperate in this regard subject to national legislation and international law in order to ensure that such rights are supportive of and do not run counter to its objectives" (Article 16.5). This conditionality has the potential for being used to reduce the scope and functioning of IPRs if adequate international pressure can be mounted. The USA refused to be a signatory of the Convention on the ground that it failed to sufficiently protect its' biotechnology industry.

(d) The Agenda 21 emphasises establishing enabling mechanisms for development of biotechnology in the developing countries:

> "There is therefore a need to strengthen the endogenous capacities of developing countries by means of new international initiatives to support research in order to speed up the development and application of both new and conventional biotechnologies to serve the needs of sustainable development at the local, national and regional levels." [Appendix 21, 16.43, See Also Article 18 of the Convention]

Much to the chagrin of the USA, Agenda 21 emphasises enhancing safety and developing international mechanisms for cooperation of application of biotechnology:

> "There is a need for further development of internationally agreed principles on risk assessment and management of all aspects of biotechnology, which should build upon those developed at the national level. Only when adequate and transparent safety and border-control procedures are in place will the community at large be able to derive maximum benefit from, and be in a much better position to accept the potential benefits and risks of, biotechnology. Several fundamental principles could underlie many of these safety procedures, including: [primary consideration of the organism, building on the principle of familiarity applied in a flexible framework and as appropriate, step by step and case by case]; complementary consideration of risk assessment and risk management, and classification into contained use of release to the environment." [16.32]

However, the text of the Biodiversity convention do not adequately reflect the bio safety and regulatory concerns. Focus of Article 14 is an environmental impact assessment of developmental projects or other dangers to biodiversity rather than problems of environmental and health security in the area of biotechnology. Further, it may also be mentioned that the convention makes no reference to the threat to biodiversity from biotechnology induced homogeneity. At the same time, it must be stated that the Article 19.3 called for the partos "to consider the need for and modalities of a protected" regarding safety procedures of biotechnology.

(e) And finally, even though the Global Environmental Facility (GEF) set up under United Nations bodies but managed by World Bank has been accepted as an interim mechanism facilitating the undesirable entry of World Bank into environmental funding, the stubborn stand of the Third World countries succeeded in including clauses in the finance chapter of Agenda 21 to democratise the structure of the GEF. The ultimate authority of the mechanism to administer the global diversity fund has been accepted to be the conference of parties which is a one country one vote body. The continuation of the GEF is conditional upon its democratic functioning.

The Clinton Administration has expressed its intention to sign the Biodiversity treaty with a "interpretative statement" that virtually repudiates the third world position: "Fair and most favourable terms of technology transfer is best determined by the market' "practical measures" for such a transfer cannot include coercions of biotechnology companies; intellectual property rights are always "supportive" and cannot "run counter" to the objectives of the convention and that there is no need for a special clause on safety. All these only underline the fact that the North-South Confrontation on biodiversity and biotechnology is likely to become even more severe in future.

Section 4: Pollution and Climate Change

Pollution has emerged as a critical threat to environment degrading soil, water and atmosphere. Evidence of adverse impact is day by day mounting: More than 1.2 billion hectares of vegetated land—an area as large as China and India putting together—is estimated to have been significantly degraded since World War II adversely affecting agricultural productivity. In the last decade per capita food production has declined in 69 developing countries. [*World Resources, 1992-93*] Freshwater scarcity is emerging as a major problem in many parts of the world. Decline in fish catches in major parts of the world from the late eighties is partly related to the growing ocean pollution and degradation of coastal systems.

It is the air pollution which has, however, caught world wide imagination. Excess particulate is a near universal urban health hazard. Acid rain and smog were the initial dramatic statements of air pollution. Evidences are mounting that the greenhouse gases are leading to an increase in global temperature. According to the most comprehensive study conducted so far by Inter governmental Panel on Climate Change (1990), mean global temperatures will rise at the rate of 0.3 degrees C per decade if current trend of emissions continue unabated and global warming will have disastrous consequences for agriculture and low lying areas. Finally, there is growing evidence of the depletion of the ozone layer in the stratosphere threatening to cause serious health hazards for the world population.

The problems created by man made changes in the composition of the atmosphere cannot be resolved by any country individually. There exists already a body of international treaties and conventions regarding air pollution. Europe has taken positive steps towards drawing up standards of "critical pollutant loads". Regional cooperation for systematic observation of air pollution, assessment and information exchange for Europe and North America has been established with the 1979 Europe Convention on Long-range Air Pollution. Two conventions on the ozone layer (1985 Vienna Convention and 1987 Montreal Protocol) were important initiatives for international action. However, the loading of the atmosphere with ozone depleting substances has continued to increase. The pressure from environmental groups have succeeded in imposing increasingly stringent controls on the emission of carbon monoxide and carbon dioxide, major components of the greenhouse gases. But it has been estimated by the IPCC that a 60 per cent reduction in carbon dioxide emissions is necessary to stabilise carbon dioxide concentration in the atmosphere at present levels.

Rio Conference sought to take forward this process of social control of atmospheric pollution through setting more stringent standards and timetables for implementation. The Convention on Climatic change, however, was bogged down in controversies. Central to any international agreement on measures to contain the adverse climatic changes is the issue of relative contribution of various country groups towards emission of green house gases and the relative adjustments each has bear. The North-South division came to the open from the above starting point of negotiations itself. The influential World Resource Institute, Washington DC in collaboration with UN agencies had held that both developed and developing countries shared the responsibility for the green house gas emissions. The net contribution of both groups of countries was stated to be nearly equal. The top six culprits accounting for 53 per cent of the global emissions included top ranking industrialised countries of USA, USSR and Japan as well as populous developing countries of China, India and Brazil. While the developed countries' emissions were mainly CO_2 and CFC gases from their industries, the emissions of the developing countries were CO_2 from deforestation and land clearing and methane from wetland agriculture. In terms of per capita emissions, the oil producing countries and some of the least developed countries also topped the rank list. [*World Resources 1992-93*, p. 210 and p. 208]

The environment groups in the south rightly condemned the above assessment as the "based less on science and more politically motivated jugglery". It was alleged by Anil Agarwal and Sunita Narain that the intention of World Resources Report seemed to be "to blame the developing countries for global warming and perpetuate the current global inequality in the use of earths environment and its resources". Apart from the weak data base for emissions for changes in the land use pattern and wet land cultivation and methodology of arriving at an index of contribution to global warming, serious ethical issues, such as the need for equitable distribution between every member of humankind, the earth's capacity to absorb emissions of green house gases, the need for accounting the historical share of green house gas emissions by various country groups etc. were also raised. Accordingly, alternative indices of green house gas emissions have been constructed a summary of which is available in World Resources 1992-93 (p. 209).

Agarwal and Narain have attempted to determine the natural sinks for carbon dioxide and methane for each country in proportion to its population. Any excess emission of green house gas beyond the quota has to be purchased from countries that have surplus quotas. According to the approach proposed by Yasumasa Fuiji, quota for carbon emission by each country is to be calculated by allocating equal carbon emission right to every person between 1800 and 2100. Unspent permits can be inherited. USA's quota would turn out to be only a tenth of its present per capita level. North would have to substantially reduce their carbon emissions while South would be able to increase their carbon emission substantially. A similar approach developed by Kririk Smith attempts to measure cumulative surviving human induced emissions of green house gas per capita for various countries. The historical pollution debt of USA is 260 mt of CO_2 per capita in contrast to 6 mt for India [World Resources 1992-93, p. 209). Fidel Castro succinctly summed up the situation: What has to be paid to the earth is the ecological debt of the rich nations and not the foreign debt of the impoverished countries!

The preamble to the Convention on Climate Change also recognised the historic responsibility of the developed countries that "the largest share of historical and current global emissions of green house gases has originated in the industrial countries, that per capita emissions of the developing countries are still relatively low and that the share of global emissions originating in developing countries will grow to meet their social and development needs".

But no developed country was willing to pay for its historic debt or agree upon an immediate programme of action to help island economies threatened by sea-level rise or regions under threat of desertification. Many among them, notably as USA, was even unwilling to accept a concrete time table for achieving the objective of the Convention: "stabilisation of green house gas concentrations...at a level which would prevent dangerous anthropogenic interference with the climate system". To ensure stabilisation of atmospheric concentration of carbon dioxide, a significant reduction in the present level of emission is required. But the article 4 that deals with control targets does not aim any reduction of emissions but only their stabilisation at 1990 levels by the industrialised countries. Even this is worded in such a vague manner that the USA continues to dispute any binding commitment to stabilise green house gases at 1990 level by the year 2000. USA feared that any binding commitment to stabilisation of green house gas concentration or emission may reduce their economic competitiveness with Japan and Europe who in the last two decades have far outdistanced USA in energy efficiency. Finally, the only firm commitment to action that every one agreed to was a system of reporting on the progress towards stabilisation, even if they necessarily do not achieve the target. The developing countries are to be given certain grace period and funds to enter the system of reporting. Special assistance to island and desert economies were postponed. Importance of technology transfer and financial aid for the implementation of the commitments was accepted but their quantum and modalities are left to GEF and future conferences of the parties.

The Northern manoeuvres with respect to the Convention on Climate Change made it clear that they were seeking a Forest Convention as the cheapest way of cutting carbon dioxide concentration. To the North, the tropical forests were simply sinks for the carbon dioxide pollution they generate. To the developing countries, however, these forests are also source of livelihood to millions of tribals and a valuable source of resources. Led by Malaysia and India they doggedly refused to agree upon any supra-national control of forests that the North was seeking to impose through legally binding Convention on Forests. The failure of the North to offer any credible compensation mechanism such as increased foreign aid to cover the revenue that developing countries would have to forego in reserving the forests for global welfare, sealed the fate of any chances of a Convention on Forests. Instead the "Non-legally binding authoritative statement of principles for a global consensus on the management, conservation and sustainable development of all types of forests" properly known as "Forest Principles" was adopted at the Rio Conference to guide 'the management conservation and sustainable development of forests to provide for their multiple and complimentary functions and uses". Even, the then US democratic Vice President Candidate Senator AL Gore had to admit that the failure to reach a forest treaty was principally the result of Bush Administration's refusal to accept any binding commitment in the climate treaty. Any resolution on the excess emission problem that does not take into account the reality of global inequity and historical liabilities would be severely restricting the growth prospects of the developing countries.

There was also an attempt to intervene in the management of toxic chemicals and hazardous waste including the prevention of illegal international traffic in toxic and hazardous waste. The programme provided for expanding and accelerating international assessment of chemical risks, harmonisation of classification and labelling of chemicals, evolving information exchange on toxic chemicals and chemical risks, establishing a risk reduction programme, strengthening national capabilities to manage chemicals, and preventing illegal international traffic in toxic and dangerous products. With respect to hazardous wastes, Agenda 21 attempted to "within the framework of integrated life-cycle management, the overall objective is to prevent to the extent possible, and minimize, the generation of hazardous wastes, as well as to manage those wastes in such a way that they do not cause harm to health and environment." [20.6] But the control recommended over cross boundary traffic, it must be said, in hazardous waste is far from adequate.

Section 5 : Environment and Development

The deliberations on Convention on Climate Change and Forest Principles brought to the fore the potential tension between development on environment given the fact that such an integration is sought within a global order that created and perpetuated the underdevelopment. Not only was even the rhetoric of intervention to modify the institutional parameters of neocolonial international order given up but also, as we have seen, their concerted effort was to shift the burden of cleaning up the earth on to the back of the third world. Lack of sufficient sensitivity to the development concerns made even the name of the basic declaration of the conference itself a source of contention: The 'Earth Charter' was criticised as not reflective enough of development concerns and had to be renamed into 'The Rio Declaration on Environment and Development". The declaration became a distillation of the north-south divide pervading the global debate on poverty and environment. It did reflect the third world development concerns but only to be post scripted by an 'Interpretive Statements for the Record by the United States' disclaiming that development is not a right but only a goal; refusal to accept any implication of liabilities or obligations for the environmental change, assertion that trade measures may be used to protect environment etc. The Conference could not even extract a commitment to the target of 0.7 per cent of donor country GDP—a target that UN system had accepted long back. In the place of estimated cost of $125 billion for the implementation of Agenda 21, additional aid pledged totalled to a mere $2.5 billion. Even the meagre additionalities was conditional to being routed through GEF, with substantial control by World Bank, a non UN body.

The above international situation have to an extent strengthened the tendency in certain quarters to look upon with suspicion any attempt to integrate environment in the development decision making process even within the framework of national priorities. Environment concern is viewed as yet another additional constraint on economic growth.

We believe such a mode of thinking to be a short termed perspective of economic growth. Before further discussing this point we wish to emphasise the importance of economic growth in the development process. Given the magnitude of absolute poverty in Third World countries, a significant acceleration in economic growth is undoubtedly a pre condition for development.

Not only must there be an acceleration in economic growth but also it should be sustainable in the long run. Cutting down entire stock of trees in the country for export in a frenzy of economic growth may enable an immediate quantum leap in the rate of economic growth. But the moment the stock is depleted not only would the income flow stop but also the unforeseen chain reactions on soil erosion, water cycle, micro climate, etc. would have disastrous consequences in production in other sectors such as agriculture, fishing, etc. The short term economic leap only impoverishes the future. Integration of environmental concerns in the economic growth process precisely seeks to avoid such self defeats and to ensure long run viability of the acceleration in economic growth. Of particular importance is the need for "integrated approach to the planning and management of land resources" (Agenda 21, Ch. 10), coastal and marine areas (*Ibid.*, Ch. 17), forests (*Ibid.*, Ch. 11) and freshwater resources (*Ibid.*, Ch. 18). Viewed in this fashion such an integration is not an additional constraint but a precondition for development. The present developed countries could wantonly disregard the environmental considerations in the growth process because of the advantages conferred on them by the international framework of colonialism and neocolonialism that characterised their economic transition and maturity.

Dissatisfaction with unsustainable pattern of Western production and consumption which is wasteful and environmentally degrading and in the long run endangering the existence of humankind itself gave birth to the concept of sustainable development. It has had varied interpretations but has gained widespread acceptance as a powerful metaphor that emphasises the environmental linkages of the development processes.

The concept of sustainable development has today acquired also social dimensions. If economic growth is to be characterised as development it should lead to the improvement of the standard of living of the majority of the people. In the present day context of developing countries, it means at least that the basic needs of the population such as minimum food, drinking water, shelter, education, health care, sanitation and recreation should be met. Agenda 21 gives substantial importance to "combating poverty" (Ch. 3), "protecting and promoting human health conditions" (Ch. 6) and "promoting sustainable human settlement development" (Ch. 7). In short, economic growth should not only be environmentally sustainable but also socially equitable.

Explicit introduction of long run and indirect effects of man-nature interaction and social aspects of distribution in the economic growth process necessitates social regulation of the same. Priorities would have to be set and resources guided accordingly. This in the broad sense is what is known as planning. But this planning can be effective only if it is decentralised and participatory in nature. Though the concept of planning is an anathema to the Agenda 21, given the obsession with free trade policy, it does emphasise the importance of local authorities (Ch. 28) and the need for strengthening of the role of workers and trade unions (Ch. 29), farmers and their organisations (Ch. 22), youth (Ch. 25), indigenous people (Ch. 26) and non governmental organisations (Ch. 26).

Thus we would define sustainable development to be environmentally sustainable, socially equitable and planned and participatory rapid economic growth process. So defined, its component elements could play as powerful slogans for mass mobilisation for social change. In the contemporary developing countries, the nature of capitalist transition is such that it condemns the vast majority of the people permanently to the informal sectors deprived of even modicum benefits of economic growth. Movements for basic needs could be a powerful mode of mobilisation of these vast masses. Similarly, ecological struggles can be an important means of mass mobilisation and conscientisation for social transformation. However, these movements that are today emerging throughout the Third World have got to be consciously linked to a platform that challenges the new world economic order that is throttling conditions for self reliant economic growth in the Third World countries.

5

ECO-POLITICS
A New Trend in Global Environment Degradation

HEMA YADAV

Eco-Politics

THE term Eco-Politics implies the political wrangling centred on ecological problems in the modern world. It is the conflict between North and South over the issue of environmental protection. The politics is that of blaming each other for ecological destruction that has been done by mankind to pursue economic development,

The current environmental problems are global in nature. Over the past few decades fundamental changes in world economy have disrupted the ecological equilibrium which is conspicuous in the trends of environmental degradation.

The global environmental problems as such are becoming so alarming that they demand immediate attention of the world governments and environmental activists. But it is unfortunate that every government or agency is viewing the environmental problems from their own perspectives, which is solely committed to further one's own interest. As a result the same problem is invoking different reactions from different corners. The Governments of various countries having similar backgrounds and targets have polarized and they collectively lobby for their group interest. The state of affair is so chaotic and pathetic that the remedies for environmental problems have taken back seat.The objective evaluation of the magnitude of environmental problems *viz.*, identification of its root cause and a strategy for overcoming them has been replaced by subjective consideration.

These environmental issues which were earlier considered as national issues have now taken global dimensions. As a result the world has polarized in two blocks. The developed world consists of affluent G-7 viz., USA, UK, France, Germany, Japan, Canada, Sweden and developing countries of southern hemisphere have grouped as G-77. They are also called as North and South blocks respectively on the basis of their geographical location.

The crux of the problem is simply to identify environmental problems and to overcome them. The ecological-politics has gained momentum since Stockholm conference. The approach of national blocks is to interpret the problem in such a manner that their group interests are safeguarded and remedial measures are prescribed at the cost of the others. As a result, North block brands the South as a culprit in creating environmental degradation whereas the South condemns the North as a villain responsible for the whole tragedy (*The Hindu*, 1992)

The environmental problems, their approaches and deliberations at various conferences and summits have clearly exposed that the ecological issues are the recent pitch of political consideration. In fact Eco-politics is the neo-imperialism in the grab of ecological considerations.

Review of the Problems

The discernable trend was witnessed in the Stockholm conference (1972), where a lengthy list of environmental issues had been highlighted. The conference raised the issues of pollution in the developed world and that of resource scarcity in developing world (Sharma, 1990). It was this conference which brought the words like Ecology and Environment in the common parlance of development. In the period between 1972 to 1987, some progress was made where governments and planners become environment conscious to a certain level, but in general terms the global environment continued its downward escalation. In 1987, the eco-politics began to take fore seat with the publication of Brundtland report (UNCED, 1987). The chief contribution of this report was to stimulate a debate on translate the concept of sustainable development into practical terms. In December 1989, the United Nations called for a world conference to discuss the basis of sustainable development, check further environmental degradation, repair damage already done and to set the process of reversing the calamitous current trend (*The Hindu*, 1991). The Earth summit was the outcome.

The Earth Summit and Eco-Politics

As the new world was balancing in terms with geo-political and economic realities, with unification of Germany and disintegration of USSR, the Earth Summit under auspice of the UN was held at Rio De Janeiro from 2nd to 12th of June, 1992. The idea of holding the Earth Summit was to focus on environmental issues threatening the future of our planet. For two years prior to 1992 several pre-Rio conferences to chalk out the agendas for Earth summit were held at Nairobi, Kaulalampur and Bejing. The heads of more than 150 nations and over 30,000 delegates from all over the world gathered to discuss the fate of the Earth. The Agenda 21 or the Rio declaration on the principles of General Rights and Obligations were adopted by the head of the nations. However, unfortunately the Earth Summit turned into North South confrontation.

The Developed nations of North, particularly the USA, in spite of her present trend of overconsumption of earth's resources and reckless pollution of environment refused to accept their share of responsibility. The developing nations of South on the other hand unitedly refused to succumb to pressure from North to bargain away their sovereignty over forest and bio-resources. Nothing can be more tragic than the fact that at Rio the world leaders were more interested in greenbacks than green fields (*India Today*, 1992).

At each stage of Earth Summit the role played by Eco-politics was discernable. Even the Agenda 21 was diluted to suit North's reservation. A serious rift grew between the Developed and Developing nations over the question of environmental protection and suddenly the Environment became a major foreign policy issue (*The Statesman*, 1992). The game of ecopolitics as played by the USA was to make environmental protection as another instrument to dictate economic and social policies of Third World.

The Earth summit has put up questions before human kind—who is to be blamed primarily for damaging the Earth and who will share responsibility for cleaning up operation.

The major issues which led to North South confrontation were:

The Biodiversity Convention

Tropical forests harbouring the richest bio diversity have received considerable attention. The FAO (1980) studies indicate that in Asia and Pacific region alone 12 million hectares of forest are being lost annually due to logging, overgrazing and over exploitation. Under severe pressure of deforestation of prime habitat every year 100 more species enter into endangered list. About one quarter of the Earth's species risk extinction within next 30 years, i.e., 25,000 species and 1000 vertebrate species are threatened. As much as two-third of all teresstrial species are located in Developing countries.

The Biodiversity treaty for the protection of plants and animal species from extinction was the biggest disappointment. The North Nations, whose pharmaceutical and chemical industries thrive on genetic stocks of South, wants the south nations to conserve their forest and insisted on making the Tropical ecosystem to be made common world heritage. At the same time, USA at Rio maintained the view that since biotechnology is in private hands, the research emanating from original biodiversity resources has to be protected by Intellectual Property Rights of GATT.

The globalization of their forest was seriously objected by South nations. For Third world people forest are the source of livelihood. The contention of South was that North must share their profits in biotechnology and pay for forest working as natural sink.

The Treaty for Reduction of Green House Gas Emission

At Rio, the climate change convention too remained undecided under eco-political stress. The South nations were reluctant to any cut in their own carbon emission. Their argument was that the 20 per cent of population living in countries of north emit 80 per cent of global green house gas.

The American President George Bush refused to any commitment to reduce the greenhouse gas emission as it would hinder their economic development. At Rio, he announced that he would not commit the country to anything that could cost jobs and affect the life styles.

The comparative figures of CO_2 emission show that the north nations are responsible for as much as two-third of total emission. In fact, the developed nations owe the carbon debt to south nations.

Technology Transfer

Transfer of ecologically benign and environmentally safer technologies of economic development from the rich developed nations to the poor underdeveloped nations was another ecopolitical issue at stake. The world urgently needs safer substitute for CFC as it is destroying Ozone layer. The north nations produce and consume 98 per cent of CFC while the south consume only 2 per cent and they are forced to phase out CFC.

The North nations believe the technology development is commercial and those countries who want to utilise it must pay for it. The contention of the south is that north nations have created the environmental problems and must feel obliged and indebted to transfer better technologies at nominal cost.

The Fund for Environment Protection

The question who shall pay for cleaning up action of the world's polluted environment and repair of the damaged earth was a subject of great debate among north and south nations.

Holding the developed countries solely responsible for the present global environmental condition, Sri Kamal Nath, the then Minister for Environment and Forest, Government of India, said that the developing countries are now looking at them to provide "new additional and incremental" funds to take corrective measures (*India Today*, 1992). The developing nations insisted that a funding mechanism should be set up to be administered "democratically". The UNCED agenda 21 estimated $ 625 billion annually for the cost of environment clean up and repair damaged earth. As the south already spends $ 200 billion on such activity, the UNCED expected the rich nations to chip the remaining $425 billion. The rich nations, e.g., Germany, Japan, France and UK have a commitment at Rio to keep 0.7 per cent of GDP aside for assistance to South (*India Today*, 1992). However the USA offered only $150 billion to protect the world's forest. Nevertheless the financial commitment by rich nations is too small in view of the "goods and services" they have overdrawn from the environment.

The Conflict between Overconsumption and Overpopulation

The affluent, extravagant lifestyle of North have done irreparable damage to environment. The north has always blamed south nations for its increasing population and poverty. While the problem of south are serious and should be checked immediately, one cannot ignore that the north with 25 per cent of population consumes 85 per cent of the world resource.

Consider the following facts (Roy, 1992)

A child born today in a typical middle class American home in his life time will

* Go through 12,000 paper grocery bags
* Wear and discard 250 shirts and 15 pairs of shoes
* Purchase a dozen automobiles
* Go through 28,627 aluminum beverage cans
* Consume enough commercially produced electricity to burn 16,160 pounds of coal
* Burn out 750 light bulbs

At Rio the South nations pointed out it is the overconsumption and not the overpopulation which is causing the problems. In USA the world's premier consuming society people own twice as many cars and use 21 times as much plastic than people living in third world countries (*The Statesman*, 1992). According to UN data 1991 the USA with 5 per cent of the world population consume 25 per cent of the fossil fuel and disproportionate share of other raw materials. India on the other hand has 20 per cent of global population but uses only 2 per cent of world's energy .The north uses up to 1000 litres of fuel oil per head per day whereas the south can manage with 10 litres/head/day. There is something to be learnt from South (UN, 1992).

Lifestyle of north indeed depend on the resources brought from the world of poor while we in south do know that poverty is detrimental but it is not the end of the world and its wealth. However, the real culprits are the north nations who are overexploiting the natural resources to support their lifestyle.

To Save Our Planet

The ecopolitical divide between north and south is unpragmatic because the problems are no longer regional or national; they are global and are interlinked. All the nations of the world must rise above the political prejudices and self interest. The answer lies in adopting sustainable development, which should integrate development with conservation. The north and south should enter into global partnership for development which is ecologically sustainable, economically viable and socially acceptable.

REFERENCES

India Today, 1992: June 15th issue
Roy, B. 1992. North not South is Destroying Earth, *The Times of India*, March 29th.
Sharma, S., 1990. Environment—What Next? January, *The Hindustan Times*.
The Hindu, 1991. For a Global Partnership on Environment, Nov. 14th issue.
The Hindu, 1992. Hurdles on the Road to Rio, June 7th issue.
The Hindu, 1992. It's Time to Pay Up, June 14th issue.
The Hindu, 1992. For a Global Partnership on Environment, November 17th issue.
The Statesman, 1992. June 15th and June 27th issues.
UNCED, 1987. Our Common Future, Oxford.
UN, 1992. The State of The World, 1991, 1992.

6

APPLICATION OF SUSTAINABLE DEVELOPMENT CONCEPT
Some Issues

SRIKUMAR CHATTOPADHYAY

Introduction

THE Earth Summit held at Rio in 1992 reemphasised the need for sustainable development, a term perhaps first used at the time of Cocoyoc declaration in 1970 and gained popular currency through IUCN report (1980, 1990) and WCED report (1987)—Our Common Future.

The term Sustainability and its corollary sustainable development are being widely used in government policy pronouncements, scientific articles, publications of international aid agencies, professional literatures, seminars, symposia and popular media.

While unsustainability of the ongoing development processes is well manifested through environmental problems, declining productivity and such other indicators, how sustainability can be achieved is not very clear. Moreover, there is little unanimity in comprehending this term. It conveys different meaning to different people. In global context, view point of the developing countries differ considerably from that of the developed countries as evidenced during the Rio summit.

The developing countries are mainly concerned to tackle the immediate economic problems and to eliminate poverty, whereas, the developed countries' thrust is to combat industry related pollution and to maintain the already attained economic level. Various research reports suggest that the developed countries consume resources at a disproportionately high rate when compared to their resources base. It is well documented that growth of many national economies can be sustained only by continuous imports from elsewhere.

This chapter considers sustainable development in the context of developing countries only, which depend on the land, water and biomass resources.

What is really meant by sustainable development and how to apply this concept in operational terms are the points of discussion in this chapter.

Sustainable Development (SD)—The Concept

Since 1980's, particularly after publication of the report—Our Common Future—a good many literatures are devoted to deal with the concept of sustainable development and its possible application (Barbier, 1987; Brown, *et al.*, 1987; Daly, 1989; Lele, 1989; Pezzy, 1989; Redclift, 1987; Rees, 1988 among others). The present author has listed as many as 33 definitions by scanning available literatures published during 3 years from 1987 to 1989 (Chattopadhyay and Carpenter, 1990). Brown *et al.* (1987) and Lele (1989) summarized the essential elements covered in various deliberations. In broad terms these essential elements or objectives and premises cover all the following aspects.

1. Survival of human beings.
2. Survival of all other life forms.
3. Satisfaction of basic human needs.
4. Maintenance of bio-physical productivity.
5. Economic efficiency and growth.
6. Preservation of environmental quality and ecosystem.
7. Inter and intra generational equity.
8. Social justice.
9. Self-reliance and people's participation.
10. Stabilization of human population.
11. Promotion of values and ethics.

There are also attempts to cover a good many of these aspects into a single framework. As a result, instead of being precise, the definition tends to be vague and subjected to misinterpretation. So great is the confusion, that, even this vagueness is justified to be its strength by some authors (Daly, 1989) and some others (Lele, 1989) opined that this lack of precision in definition leads to the faltering of SD in operational terms.

Without going into the merits and demerits of this problem, what is significant in this wide ranging debate is that there is a growing realization about:

1. Excessive stress on bio-physical environment for short term gain is ultimately counter productive, that is there is a limit to which productivity can be raised.
2. Human survival can only be ensured by preserving the basic resource base.
3. Even the renewable resources are not perpetual unless prudent management practices is followed
4. Conventional economics does not necessarily indicate the actual status of resource utilisation.

The matter of social justice and equity are prime objectives of any form of development, so far, conceived. Quite naturally, they are also professed goals of sustainable development.

Sustainable Development Aims Goal Maximisation

Considering all the above aspects SD at the conceptual level can be assumed to maximise three goals: Biophysical, Economic and social (Fig. 6.1). Elaborating the figure further it is hypothesised that during pre Marxian economics it is the economy which used to determine the decision making process. With the advent of Marxian economics social aspects are given equal importance. Decision making processes attempt to consider not only the economic issues but also the beneficiaries. Maximum good to the maximum people is the prime objective of planning. Both economic and social goals are supposed to be maximised at this level. Now in view to the environmental concern, the operational part of the development process should take the biophysical variables into consideration. Therefore, SD advocates a development process that is biophysically permissible, economically viable and socially acceptable. In other words it is a process of development that should operate within a frame work bounded by the biophysical, economic and social coordinates. To operate within these three coordinates it is necessary to resort to a process of trade offs, primarily to recognise that there are certain limits within which the development process has to be functional. These limits can be broadly identified as Biophysical and Ethico—social limits (Daly 1989).

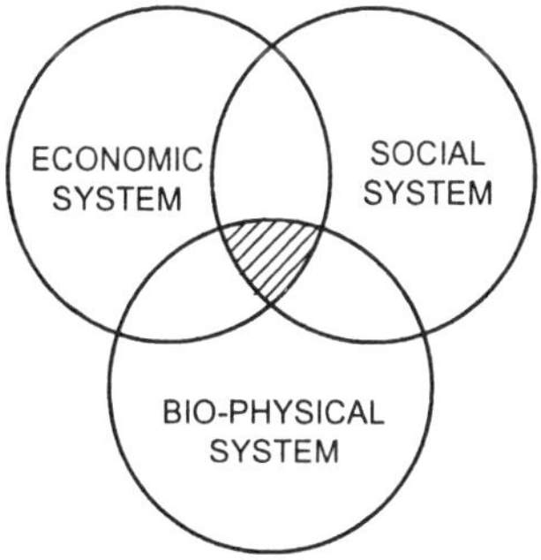

FIG. 6.1: Sustainable development aims to maximise the goals across three systems (Modified after Barbier, 1987)

The Bio-physical limit, which actually controls resource availability, resource use and economic viability is due to:

(i) finitude
(ii) entropy
(iii) complex ecological interdependence (CEI) and
(iv) resilience or self ameliorating capacity.

Apart from these four, photosynthetic limit is another factor to impose restrictions on crop/biomass productivity.

The ethico-social limit is equally important in the matter of achieving sustainability. This dimension is supposed to provide philosophical back drop of a society including the legal, executional and policy making authorities. This aspect needs much higher attention in today's context as the existing development process promotes wasteful consumption. Intergenerational equity and distributive justice in the case of present generation ought to be considered as in-built system of the development process.

The next point merits clarification is that it is the development (rather than environment of a particular ecosystem or a particular economy) which is to be sustained.

In the light of above discussion on conceptual frame the following points need to be considered as guiding principles to look for an application model.

1. The basic needs of the present society must be met and the resource base should be mentioned for the future generations (WCED, 1987).

2. In operational terms SD is a production management system in which inputs do not adversely affect the bio physical potentially of the system including its residence limit.
3. As sustainability is basically a constraint, the approach for achieving SD is necessarily interventionist.
4. As SD is conceived for the (Resource) system as a whole it is imperative to follow a system approach.
5. SD has to be pursued on the basis of renewable resource.

Applying the Concept

Considering the fact that sustainable development is accepted as the societal goal, the question is how to put the concept into practice. As it has been indicated earlier that the application of SD requires deliberate planning i.e., interventionist in approach. The interventions are warranted at the bio-physical front or resource use level and also at the socio political front or policy level. Under these circumstances, the concept of Filters—Ecologic, Economic and Social—has been advocated. A brief discussion is provided here to clarify the concept. By filter what is meant is a type of screen through which only desirable elements can pass through.

Ecological Filter

It is well established that each parcel of land is unique in character and has a definite production potential. Even desert has its own production potential. It is possible to characterise each unit of land through landscape ecological analysis covering topography, soil type, drainage condition, state of water availability and over all environmental capacity. The suitable landuse type under each of these landscape ecological units can be worked out on the basis of existing body of knowledge. Traditional wisdom helps greatly at this juncture.

A broad apportioning of area for various uses can be arrived by classifying the land. A list of landuse practices may be worked out for each of the identified landscape ecological units. Nature generates complex configuration, therefore, diversity in landuse is normal. As human prefers simple system there is a normal tendency to resort to monoculture landuse. Landscape ecological analysis provides the base for understanding the natural configuration prevalent in the concerned area before human induced simple system took place. An emphasis on landscape ecology is particularly required since most threats to a sustainable biosphere involve changes in landuse at large scale (O'Neill *et al.* 1995).

Further analysis of crop ecology and plant ecology can fine tune the ecological filter. As vegetation has a wide range of adaptability a number of crops can be raised in same type of land. To workout this ecological filter the activities suggested are Investigation, Assessment and Allocation. Investigation helps to characterise the unit/system i.e., generates raw data to define unit/system boundary. Potential use can be evolved through assessment. Allocation is an exercise of ordering the uses based on suitability/viability.

Application of ecological filter begins with Landscape ecological analysis of the concerned area. Remote sensing and other time tested field survey methods are essential tools for this activity. Selection of the method depends upon scale of activity macro, meso/regional or micro/ local.

Ecomonic Filter

Once a set of landuse or crop is found to be ecologically viable, economic consideration becomes important to choose among them. Due to multiple usability of landscape a number of landuse practices may be ecologically just but may not be economic at all during the given time point. While ecological filter is not usually bound by time factor in a normal human scale, the economic filter is time dependent as it is subjected to technological development. Efficiency, growth and stability are prime concerns under this exercise. Net internal return, cost benefit analysis, input output ratio, market value etc. are usable methods for functioning of this filter. This can be applied for each of the uses ordered through allocation under ecologic filter. In case of agriculture, individual crops may be taken into account to begin with and subsequently all the crops raised within an agricultural calendar year need to be counted. All economic analysis must internalise all possible externalities, while judging a specific use/crop practice. All three operations of Investigation, Assessment and Allocation are part of this exercise also.

Social Filter

Social acceptability is as much important as ecologic and economic criteria. Social viability of a use/project can be assessed from the standpoint of equity, empowerment, cultural tradition/diversity and availability of social institutions. Functioning of social institutions vis a vis landuse is an important index to understand working of social filter.

Adaptation to new technology by any particular group depends, to a large extent, upon the group's capacity to culturally assimilate the change. While discussing the transformation of environment Redclift (1987) concluded that there is a need to be more aware of the relationship linking people's understanding of the environment with their behaviour. Livelihood strategies that people used to pursue in ecologically fragile areas provide important clues to the kind of technologies that are most appropriate on social as well as ecological grounds.

Hence a sustainable landuse pattern can not be designed without considering social filter. Like economic filter, social filter is also subjected to temporal change. Investigation, Assessment and Allocation are necessary for activising this filter.

As ecological filter is basically nature-determined, it is imperative to understand the nature's law through ecosystem analysis in order to prepare guidelines for economic and social interventions. Once this exercise of applying filter is completed three other steps of Action planning, Implementation and Monitoring are involved in chalking out sustainable landuse pattern. These three activities form part of the filtering system as a whole, but start functioning from the end point of social filter. It is here that the People's participation is considered to be integral part of the entire process. A diagrammatic representation of this filtering concept is provided in the Fig. 6.2.

Operational Activities

Some of the operational activities to be taken up under this conceptual frame are given below.

1. Environmental Appraisal
2. Cumulative Environmental Impact Analysis and Trade Off
3. Natural Resource Accounting

1. *Environmental Appraisal*

Environmental appraisal is an assessment of the existing conditions based on which decision on desirable activities can be taken up. It is necessary to investigate the governing principles of production potential of each unit of land to act upon. Regressing the existing activities/use with the potentiality it will be possible to understand the problems,

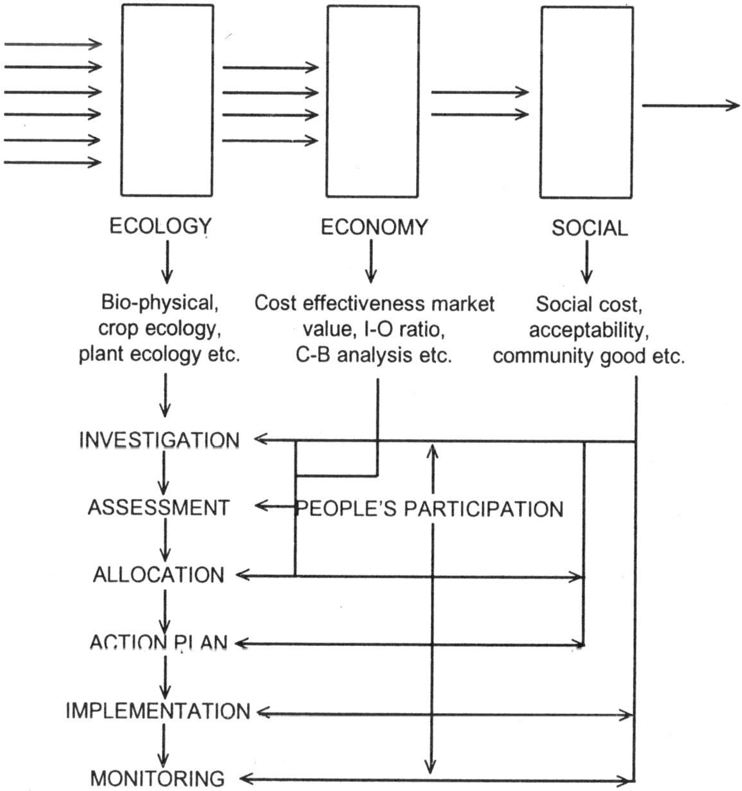

FIG. 6.2: Filtering concept for sustainable landuse system

associated causes and the possible way out. In this case problem provides the key to solution. Table 6.1 provides a sample environmental appraisal method.

2. *Cumulative Environmental Impact Analysis and Trade Off*

Impetus behind the call for sustainable development is common concern about cumulative environmental and social effects of human activities at all spatial scales. Planning for sustainable development will obviously require systematic identification and monitoring of cumulative negative trends in significant environmental variables (Rees, 1988). As these effects will be at the regional/local scale, it will be necessary to look into those problems according to functional regions based on such ecological criteria as climate and vegetation pattern, soil types and watershed boundaries. These natural regions or units ensures diversified production array (Sinha, 1990). To decide upon certain activities a trade off analysis will be most suitable, specially, when a choice is to be made.

TABLE 6.1: Environmental appraisal for landuse planning—a sample for selected land units in Kerala

Unit name	*Slope (%)*	*Soil type*	*Water availability*	*Erosion rate*	*Present Landuse*	*Desired Landuse*	*Remarks*
C1	< 05	S1	High	Low	Co, P,	Co, P, V	Protection of wetland, Dense settlement
F1	< 05	S2	High	Low	P, B, Co	P, B, V	Protection of lowland
D1	< 15	S3	Medium	Medium	Co, Mt, R	Co, Mt	Soil & Water Conservation, Dense settlement
D4	> 35	S4	Low	High	Pt, Pc, R Co, Pcm, F	Pt, Pc, F Pcm	Protection of forest, Soil & Water conservation, Scattered settlement

Note : *Unit:* C1—Coastal plain, F1—Flood plain, D1—Low rolling terrain and D4—Highly undulating hilly terrain
Soil: S1—Coastal alluvium, S2—Riverine alluvium, S3—Lateritic soil and S4—Residuum and variable shallow soil
Landuse: Co—Coconut, P—Paddy, B—Banana, Mt—Mixed tree crops, R—Rubber, V—Vegetables, Pt—Tea plantation, Pc—Coffee plantation, Pcm—Cardamom plantation and F—Forest

As an example low land conversion in Kerala can be a point of discussion. Conversion of paddy fields to coconut groves, other tree crops and settlements is widespread in Kerala. In past 30 years as much as 25 per cent of lowlands have been put to other uses. One of the common practices in Kerala is to divert paddy fields to coconut groves. Let us first consider positive aspect of this change in landuse.

The benefits are: (1) higher yield per unit land, (2) low management cost, (3) increase in production of industrial raw materials, and (4) relatively less chance of crop failure.

The negative aspects of this transformation are, (1) reduction in water spread area (lowlying areas accommodate rain water during monsoon) and thereby contribute to flood problems, (2) decrease in agricultural employment, and (3) decrease in rice production.

Through cumulative environmental impact analysis impact of land conversion can be brought out and to make a choice between paddy and coconut the following issues warrants consideration along with the negative and positive aspects mentioned above for trade off analysis.

1. Can coconut production compensate for loss due to flood
2. Can the surplus labour be absorbed in local coir industry
3. Can the demand of rice be met from external sources and what will be the additional cost.

3. *Natural Resource Accounting*

This is another aspect that need immediate attention. Few examples are available at present (Repetto *et al.*, 1988), The rationale behind this model is that Gross National Product (GNP) dose not take into account the depletion of natural stocks like forest, soil, water quality etc. All man made assets are discounted over time. This is not applied in case of natural resources. Economic loss due to loss of productivity or depletion of stock can be brought through this analysis. An example of productivity loss in Kerala is provided here.

Tapioca is one of the important crops cultivated in Kerala. It used to be considered as non responsive crop, i.e., it can be raised without adding fertiliser. Therefore production of tapioca entirely depended on natural fertility of soil. Decline in productivity is directly related to loss of soil fertility. To understand loss of soil fertility data pertaining to rate of tapioca production for the period from 1972-73 to 1977-78 have been considered as in recent years fertiliser is applied for tapioca cultivation. It is observed that production rate has come down from 18674 kg/ha to 14787 kg/ha in a span of 5 years. The annual loss works out to be 4.2 per cent of base year's production. In other words average cost of loss of soil productivity is 4.2 per cent of annual income from tapioca cultivation. Capitalising this cost with a discount factor of 0.1 the losses in future productivity are approximately 42 per cent of annual value. This indicates that 42 per cent of future income is sacrificed to have present production. From another angle society can spend this amount to stabilise productivity. Similarly, it is possible to take into account the depletion of natural stocks.

Apart from these ecologic-economic relationship and energy budgeting can be taken up towards operationalising the concept.

4. *Socio-Political front or Policy Level*

The intervention at the policy level calls for the following:

1. Environmental policy
2. Property rights
3. Distributive justice or social optimisation
4. Limiting unbridled consumption

Conclusion

The foregoing discussion has brought out that SD is essentially interventionist and proactive in nature. Thus a sustainable system (Landuse/Resource use/Activities) is one that can pass through three filters—ecologic, economic and social—each one of which has a definite regulatory role. A set of techniques need to be applied for activising these filters.

The paper put forward only certain points. There are other aspects to deal with. The question of political economy in today context is very important. The scale of operation is also an important parameter. As sustainable development needs deliberate planning and control in all spatial scales, it demands educational, planning and political process that are informed open and fair.

Acknowledgement

Thanks are due to the Director, Centre for Earth Science Studies for kind permission to complete this paper. Dr. K. Soman, Head RAD is thanked for providing facilities to undertake this study.

REFERENCES

Barbier, E.B., 1987. The concept of sustainable development, *Environmental Conservation*, 14(2) 101-110.

Brown, B.J., Hanson M.E., Liverman D.M., Merideth R.W. (Jr.) 1987. Global Sustainability : Towards Definition, *Environmental Management*, 11 (6), 713-19.

Chattopadhyay, S. and Carpenter, R.A., 1990. A Theoretical Framework for Sustainable Development Planning in the Context of a River Basin. Unpub. Report, East West Centre, Hawaii.

Daly, H.E., 1989. Sustainable Development: From Concept and Theory Towards Operational Principles, *Population and Development Review*, Washington DC.

International Union for Conservation of Nature (IUCN) 1980. World Conservation Strategy: Living Resources Conservation for Sustainable Development. IUCN, UNEP, WWF in collaboration with FAO and UNESCO, Switzerland.

Lele, S.M., 1989. Sustainable Development—A Critical Review, Unpublished report, East West Centre, Hawaii.

O'Neill, R.V., Hunsaker, C.T., Jones, D., Klopatek, M., Dale, V.H., Turner, M.G., Gardner, R.H. and Graham, R., 1995. Sustainability at landscape and regional scales. In Mohan Munashinghe and Walter Shearer ed. *Defining and Measuring Sustainability : The Biophysical Foundation*, The UN University and the World Bank, Washington DC.

Pezzey, J.. 1989. Economic Analysis of Sustainable Growth and Sustainable Development. W.P. No 15, Environment Department, The World Bank.

Repetto, R. and Magrath, W.B., 1988. *Natural Resourcs Accounting*, World Resource Institute, Washington DC.

Rees, W.E., 1988. A Role for Environmental Assessment in Achieving Sustainable Development. *Environmental Impact Assessment Review*, 8 (4), 273-91.

Redclift, M., 1987. *Sustainable Development : Exploring the Contradiction*. Methuen and Co. London.

Sinha, S., 1990. The Environmental Scenario, Development and the Earth Scientist. *Geographical Review of India*, 52 (1), 51-58.

World Commission on Environment and Development (WCED) 1987. *Our Common Future*, Oxford University Press.

7

LINKAGE BETWEEN POVERTY AND ENVIRONMENTAL DEGRADATION
A Profound Opinion

FIDEL MARQUEZ SANCHEZ

Population and Environment

POPULATION/Environment Analysis requires a focus which, together with demographic trends, embraces issues related to the population's production and consumption patterns, takes into account those policy measures and technological advances designed to impede ecosystem degradation, and, in general considers all aspects of the relationship, while avoiding oversimplifications.

The integration of population and environment into the strategy of changing production patterns with social equity, and therefore into the perspective of sustainable development, presupposes three basic requirements:

1. Goods and services capable of satisfying the needs of the population as a whole must be made increasingly available (notion of growth).
2. The quality of life all individuals associated with the process must be improved (notion of equity).
3. The progressive deterioration of the physical and biological foundations of the ecosystem, the basic for expanding the availability of goods and services, must be avoided so that they can be used by future generations (notion of sustainability).

It is important, also, to distinguish between natural resources and the physical infrastructure constructed by the population.

The complexity and variety of the functions which natural resources fulfil for the population, the natural resources provide the basis for society, serve as the essential source of materials and energy for human activity, and act as the sink where the population finally deposits the wastes generated by its production and consumption activities. These functions give rise to the capacity to sustain life. The study of that capacity presupposes the concept of ecosystem. It is ecosystems, composed of natural resources; physical infrastructure; population of different species and the systemic relations between them, that there diverse functions unfold, with different characteristics and intensities.

The analysis of the interactions between demographic trends and the natural resources base must be undertaken with a broader, more dynamic and detailed focus than that based on the concept of carrying capacity.

It must be broader, because the main causes of ecosystem degradation must be include. Since people throughout the world relate to the environment in different ways, considering population pressure on natural resources exclusively in terms of numbers is too simplistic to serve as the basis for an environmentally-sensitive demographic policy. It is obvious that number and the growth rate place demands on the natural resources base, but the scope of those demands is linked to, among other factors, the production and consumption patterns of the population.

Physical Infrastructure

The demand for physical infrastructure and its characteristic are directly related to the size and distribution of human settlements, and is closely linked to society's production and consumption patterns.

There are at least five areas of physical infrastructure that are important because of their relation to population and environment.

1. The entire material network which makes it possible to locate, gain access to and extract water for its subsequent treatment and distribution to the population and for agricultural and industrial use.
2. Sewage collection and treatment system.
3. System for the collection and final disposal of solid wastes.
4. Networks for the transport of persons and goods, include storage and distribution system.
5. Energy generating installations.

In these areas, the construction, maintenance and availability of physical infrastructure affects the interactions between the population and its ecosystem.

Many elements are not included the five areas mentioned above, such as housing construction, industry, economic and social services, technological capital, etc. Leaving them out does not imply that they have no relation to the population but rather, that they are less clearly and perhaps more indirectly linked to the environment, and therefore do not need to be treated specifically.

Population, Territory and the Environment

Countries can be grouped according to the way in which individuals are distributed throughout their territories. The most significant distinction is the difference between urban and rural areas. For instance, countries with high urbanization and nation with low degree of urbanization.

Rural Areas

In rural areas with populations too numerous for available resources there is a tendency to overexploit the supply of natural resources. Forest exploitation can be attributed to commercial logging and cultivation purposes which consequently led to deforestation and soil erosion. This intensive cultivation can be attributed to the international market pressure that exists in the economic world.

But in the developing countries, when the rural areas have abundant natural resources in relation to their population, exploitation does not always coincide with local demand, but is rather oriented toward other market which import materials and energy from those ecosystems. Because of that high external demand, the natural recovery capacity of resources is often overwhelmed, causing damage—at time irreparable—to those areas and their inhabitants. That problem is worse in rural areas well endowed with natural resources, chosen for large-scale commercial production and where for want of appropriate assessment-mechanization and agro-chemical products are used improperly, which ultimately leads to the rapid depletion of soils and water resources, also affecting the health of the population.

Notable inequalities in land distribution are found in older settlement zones, independently of the natural resource endowment. Resource are under utilized in some situations and overexploited in others. Problems produced there by, especially erosion and water shortages, have been attributed to an excess of population in relation to available land.

Environmental problems in rural areas no obstacle to sustainable exploitation of natural resources. That kind of exploitation is very important for changing production patterns with social equity because it can provide populations with enormous benefits. The speed of technological progresses that up till recently deteriorated the environment and now no longer do so-can increase agricultural production and productivity and raise employment and wage levels for rural populations, which could alter labour flows and redistribute population.

The Urban Environment

The relationship between urban populations and the environment demand special attention. There are at least five critical areas in which population and the environment of the large cities of developing countries intersect:

* Water.
* Air.
* Land.
* Wastes.
* Transport.

Given the current demand for water and the economic and political cost of transporting it over distances, the capacity to recycle its must be developed and consumption patterns must change. The existence of serious deficiencies in domestic drinking water supply is recognized; a large proportion of the poorest population lacks water.

The size of urban populations is an important factor in the emission of contaminant gases and particles into the air, but most research points out that the location of the cities and industries and the excessive use of motor vehicles are the main causes of serious air pollution, which is at levels hazardous to health.

With regard to land, the tendency is to consider only those problems which exist in rural areas, such as soil erosion or loss of fertility. But the territorial spread of cities and changes in land use within them are also processes which impose conflictive relations between the population and the environment.

The high growth of population in the urban peripheries, overloading the available capacity of many basic services (education, transport, health, etc.), and in some cases worsening the deficit of drinking water and garbage and sewage disposal.

With respect to wastes, the majority of the urban population in the developing countries live without sewerage to eliminate waste water, and treatment systems are virtually nonexistent. Similar problems occur in the areas of garbage collection and soil-waste treatment. Increased urban poverty and the expansion of cities into zones lacking minimum services have increased the use of rivers for the elimination of wastes, aggravating water problems. Undoubtedly, those deficiencies have played a major role in the transmission of diseases which are propagated by polluted water, such as cholera. Industrial wastes are even more hazardous.

With regard to transport, the lack of control over the number of vehicles has led to growing emissions of contaminant gases.

Poverty and Environment

In the world, these are marginalized by different causes, such as, ethnics, linguistics, age—children, young or old—, sex and physical limitations. But the greater margination is given by poverty. Such is the case, if the majority of the cases manifested the characteristics mentioned above.

The highest concentration of poverty is in the rural area of Africa and Asia, and about 50 per cent of all poor live in South Asia, where 30 per cent of the world population live.

Nevertheless, the highest concentration of poor are in Africa specially in the Sub-Saharan countries. But still, the poor of African poor are approximately only 16 per cent of all the poor people of the world. However, it must be noted that the African poor comprise 50 per cent of the total population of Africa.

The high percentage of poverty is a manifestation of lack of equity and is one of the most important political and economic challenge for the governments. But, poverty is not a new phenomenon in the world. Even in the past decades the process of concentration of the level incomes, provoked an increase in the poverty level not only in absolute figures but also in relative numbers.

This situation is fundamentally present in the developing countries (Table 7.1).

The heterogenity of poverty is manifested in the different degrees of satisfaction of the most important necessities of the poor households, in the territorial localization, in the occupational characteristics and in the different degrees of intergenerational poverty transmission.

TABLE 7.1: Global distribution of poverty (1990)

Region	*Percentage*
South Asia	49.0
South-east Asia	11.3
Middle East and North Africa	33.1
Sub-Saharan Africa	47.8
Latin America and Caribbean	22.5
East Europe	7.1

Source: The UNESCO Courier, No. 143, p. 79.

Towards the end of the 1980s, the problem of poverty gained an unusual importance. It received first priority from the international organizations, such as the United Nations, The United Nations Program for Development (UNPD), the World Bank, the Inter-American Bank for Development, and the International Monetary Fund, among others.

For a better comprehension of the great preoccupation that exists in reference to the population and poverty problems, it is necessary to say that the following are the reasons to explain the recent importance received by that phenomenon. Some of these are:

(i) The present dimensions of the socioeconomic problems at world level, specially in the underdeveloped countries.

(ii) The comprehension that poverty supposes a situation of social injustice that could contribute to national and international instability.

(iii) The organization with enough coherence of new approaches about development problems in which social aspects will acquire a greater importance.

(iv) The reappearance of North-South contradictions as a significant element of the system for the economic and political international relation after the collapse of the former socialist states and the cessation of the East-West contradictions, among others.

Of course, it is not right not to mention the opinion of South Commission related to the development of human resources.

The Commission gives the following opinions:

"A firm determination to develop human resources indicates the need to systematically evaluate the government politics in function of objectives and social achievements. The growth of Gross Domestic

Product as the main objective and it is usage as the most important development index, request the formulation and evaluation of development policy that should serve as guide for a wide range complex of social and economics indicators, which considers the public welfare and the human development in the widest interpretation". (South Commission 1991, p. 116)

The objectives that the Commission is asking for, considering other kind of development indicators, are as fellow:

(i) To make easy mobilization of poor people by increasing their participation in the administration of main services.
(ii) To give all countries the necessary instruments for increasing efficacy of their socio-economics and environmental politics.
(iii) To question the terms already used in international forums and conferences about development.
(iv) To explore new ways of development, because the old ones are proven to be not adaptable to Southern countries.
(v) And increase in the negotiating power of underdeveloped countries to counterarrest the facts of adjustment policies.

In accordance with the Commission, there are three elements in the policies which are not favourable and adequate to human development.

— Lack of systems of social welfare to provide benefits to the most disadvantaged people (deficiency in health, education, water and sanitary services).
— Unequal distribution of wastes of social welfare in detriment to the most impoverished people, benefits the high and medium income families.
— Rejecting to traditional acknowledgement as consequence of bad-understanding of "modernization".

It is convenient to make a brief synthesis about main indications of United Nations Programme for Development because it is the international organizations with the most amplified opinions in diagnosis and propositions to shoot the actual poverty levels in the world.

In the report about Human Development, United Nations Programme for Development have introduced redefinitions in concepts and methodologies to identify the poverty in the world, and have arrived at the conclusion that the indicator of "per capita" incomes is not a sufficient condition to classify the poverty. This therefore lead to the development of the "Human Development Index" which combines socio-economic indicators as:

* Life Expectancy at Birth.
* Education.
* Incomes.

In the 1992 Human Development Report, United Nations Program for Development arrived at the following conclusions:

— The growth in Economy is not automatically translated to better level of life, neither national or international level.
— In the international market rich and poor countries have unequal competition.
— Have no free action in global markets.
— Policies to give an international network of social security for poor countries and persons are needed.
— The world should design a new agreement to ensure a sustainable human development in a full peace world.
— Main conclusion, in a quantitative plane, 60 per cent of the world population receives 5.6 per cent of richness while 20 per cent richer concentrate about 82.7 per cent of richness.

According to the Human Development Report of 1994, the distance between rich and poor is increasing during the period 1960-1991 (see Fig. 7.1).

To get better results in fighting against poverty the United Nations Program for Development propositions are:

— To reduce military charges in all nations, to generate an international saving to permit the biggest coverture to programs and help systems for development.
— A new system of Official Help for Development where at least, two thirds (2/3) goes to most poor countries and shall be designed to have more than 20 per cent from human development.
— A need to restructure of external debts to profit non-development countries.
— A tax system to generate a transference of resources from rich to poor countries.

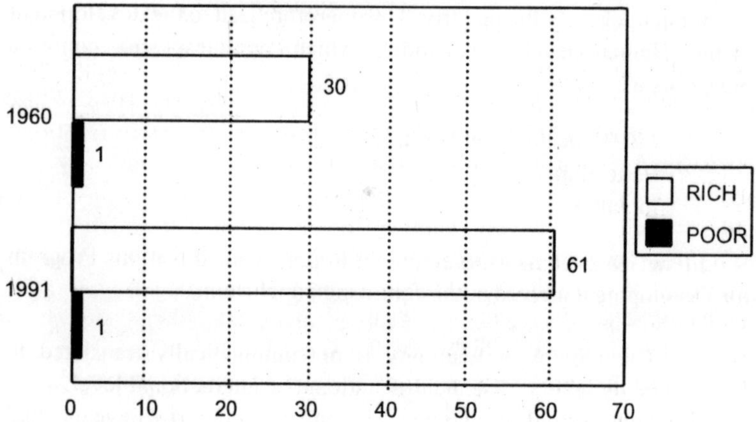

FIG. 7.1 : Increasing distance between rich and poor (Percentage in income distribution)

Source: Human Development Report 1994, UNDP, p. 40.

— Elimination of protectionism of developed countries in international market, specially in textile and agriculture sectors. (UNDP 1992)

In the developing countries annual growth now is posted at 1.94 per cent. Furthermore, this growth of population is marked by an increase of big cities. For instance, in developed north, 77 per cent of the total population lived in urban areas while in Latin America there is about 72 per cent.

In 1950 Latin America was considered an essentially rural and agricultural continent, and in 1980 the panorama had change due to the presence of the majority of the latinamericans in the great metropolis.

This situation is due to the high incidence of migration from the rural areas to the city. This relevant migratory process brought about important transformations in the structure of the labour market and, consequently, of the employment.

In the Fig. 7.2, you can appreciate the decline of the population employed in agriculture: from 55 per cent in 1950, to 32 per cent in 1980 and about 26 per cent in the following decade. At the same time, an increased the participation in the industry and service sectors, can be observed especially in the service sector. It must be taken into consideration that for the period 1950-1990, which is the period under study, an increase of the employed in the service sector was posted at 16 per cent for the period 1950-1980 and about 6 per cent for the period 1980-1990. This means that in the last year practically the 50 per cent of the total of the employed belonged to the service sector.

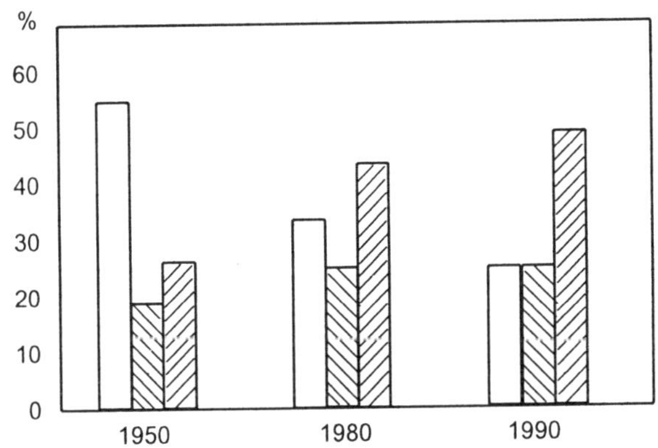

	1950	1980	1990
AGRICULTURE	55	32	26
INDUSTRY*	19	26	26
SERVICES**	26	42	48

FIG. 7.2 : Latin America: Structure of employment 1950-1990

Source: Elaborated by author whit Data of ERPLAC 1982 and ECLAC 1990
* Include Mining, Industry, Construction and Electricity.
* Include Trade, Transport and Service.

It is worth mentioning that the same is the result, among other things, of the movement of the labour force from the agriculture to the industry due to the model of imports substitution, but since enough jobs were not produce, the increment of hand work supply could not be absorbed in the cities, which was marked by the demographic growth and the strong migration from the country to the city.

On the other hand, the tendency of poverty increment in the urban areas kept up and even accelerated more with an increment in the 1980s of 9 percentage points, passing from 30 per cent in 1980 to 39 per cent in 1990. This meant in absolute numbers, and entrance to the army of the poor of 52.6 millions of inhabitants of Latin America region, and in the rural zones it increased in 1 per cent in equal period.

The absolute numbers can be observed in the Table 7.2.

TABLE 7.2: Persons considered poor or in extreme poverty in Latin America

(in millions)

Years	*Poor*	*Urban*	*Rural*	*Extreme Poverty*	*Urban*	*Rural*
1970	112.8	41.6	71.2	60.0	18.7	41.3
1980	135.9	62.9	73.0	62.4	22.5	39.9
1986	170.2	94.4	75.8	81.4	35.8	45.6
1990	195.9	115.5	80.4	93.5	44.9	48.6

Source: ECLAC, 1993.

It is necessary to point out that the problem of the poverty transcend from being only a rural problem to urban phenomenon. We can observe that in 1970, the number of poor urban represented only 36.9 per cent of the total poor of the region while in 1990, 59 per cent lived in the urban zones, mainly in the big metropolis of the region.

Developing Countries, Environment and Poverty in Figures

Developing Countries:

— Almost a third of the population, about 1.3 billion people, live below the poverty line.

— About 200 million people are severely affected by decertification.

— Every year, some 20 million hectares of tropical forest are cleared outright or grossly degraded.

Arab States:

— About 73 million people still lived below the poverty line in 1990, and more than 10 million people were underfill.

— With less than 1000 cubic metres of water per capita available each year, about 55 per cent of the people experience serious water scarcity.

East Asia and South-East Asia and Pacific:

— Nearly 170 million people below the poverty line in 1990.

- During 1981-90 more than 3 million hectares of tropical forest were lost.
- Air pollution is a major problem in the region. In Bangkok alone, 800 motor vehicles are added each day.

Latin America and the Caribbean:

- About 200 million people were below the poverty line in 1990.
- The Gini coefficient of land distribution is higher than 0.75.
- The income share of the richest 20 per cent of the population is 15 times the share of the poorest 20 per cent in many countries.
- Pesticide consumption per thousand people is the highest among developing regions, with adverse implications for the environment.

South Asia:

- More than 560 million poor people nearly half the world's poor population.
- Every year, about 4 million hectares of land are deforested.

Sub-Sahara Africa:

- About 170 million people (nearly a third of the region's population) do not get enough to eat.
- During the past 50 years, on average 1.3 million hectares of productive land have turned into desert every year.

Conclusion

In the countries of the Third World, the quest for sustainable development is the quest for development itself. The development is not only growth, but also as the transformation of economic and social structures with the goal of raising the population's quality of life and achieving the gradual formation of new ethical values.

According to Fidel Castro, "It is precisely this process of development that has been held back in the South, not as a result of accidental events or circumstance, but because of certain type of inherent social relations and ways of organizing production. This situation of backwardness and poverty is, if you will, the most unsustainable aspect of this model of development." (Castro 1992)

Underdevelopment world poverty is closely linked to the deterioration of the environment. Dependent on economies based primarily on the exploitation of natural resources, and lacking in both the financial and technological conditions necessary to exploit these resources properly, these nations are quite literally faced with no other alternative for survival than to over exploit these resources. This mismanagement engenders even greater poverty, due to a greater lack of financial and technological resources needed to confront increasingly adverse environmental conditions.

This results in a degrading vicious circle, as evident by Fig. 7.3, caught between these two phenomena, it is precisely the resources that are the source of man's existence that are destroyed, not out of ignorance, but simply in order to survive one day more.

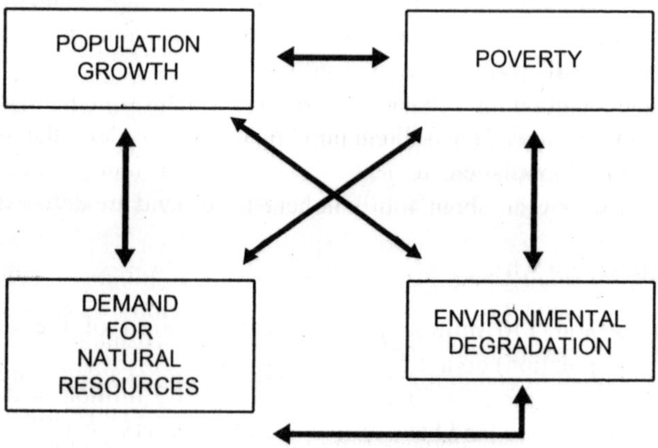

FIG. 7.3: The vicious cycle of population growth, poverty and environmental degradation

The developing countries are home to a growing number of people whose nutrition depends hypothetically on a fraction of a hectare of arable land, as a result of population growth and soil deterioration. There is no access to financial and technological resources that would allow. Third World farmers to increase productivity and crop yields to the extent necessary for maintaining sufficient production in areas already under cultivation. It is obvious that the only short-term solution available to these individuals is to subject new areas to their backward agricultural practices, the direct cause of some of the worst forms of environmental deterioration. An aggravating factor in some regions of the Third World, is the phenomenon of the displacement of individual farmers form the most productive lands by the extension of large landholding, or conversely, the repeated subdivision of land into parcels that are even more reduced and unavailable. Thus, another vicious cycle closes in around the impoverish farmers of the underdeveloped world appear to have no possible escape.

In conclusion, the conservation and environmental improvement strategies implemented by the government are considered to be the necessary steps towards eradicating the causes of the infuriating poverty that continues to ravage the Third World on the eve of the 21st century.

REFERENCES

Castro, Fidel, 1992. "Tomorrow is too late". Rio de Janeiro.

CEPAL: 1994. La cumbre social: Una visión desde Amërica Latina y el Caribe. (Nota de la Secretaria). Bases para un Consenso Latinoamericano y del Caribe sobre la Cumbre Mundial de Desarrollo Social.

ECLAC-CELADE: 1993. "Population, social equity and changing production patterns". Santiago de Chile.

Gandhi, Indira, 1972. "Life is one and the world in one", in Environment Stockholm,

Dasgupta, Partha and Karl-Göran Mäler, 1994. "Poverty, Institutions and the Environment-Resource Base". World Bank. Environment Paper Number 9, Washington.

Jagannathan, Vijay, 1989. Poverty, Public Policies and the Environment. World Bank. Environment Working Paper No. 24.

PNUD-CIDEAL, 1993. "Reforma Social y Pobreza". Revista *El Economista Mexicano.* Colegio Nacional de Economistas. Mexico.

South Commission: 1991. "Challenge for the South". Edit. Fondo de Cultura Economica, Mexico.

United Nations for Development Programme, 1993. "Human Development Report 1992". PNUD. Tercer Mundo Editores, Santafe de Bogota. Colombia.

United Nations for Development Programme, 1995. "Human Development Report 1995".

8

A NOTE ON INTEGRATING ECOSYSTEMS APPROACH INTO VILLAGE AND FARM DEVELOPMENT

K.V. SUNDARAM

Introduction

The Ecodevelopment Concept

AS we descend from the macro and meso levels to the micro level, problems become more specific demanding particularistic solutions and the context becomes more and more human centered. Thus ecological and cultural contexts assure great significance and the solution has to balance the short-term and long-term needs. The term "Ecodevelopment" is preferred in the micro level context. UNEP (1974) has defined Ecodevelopment as follows:

> "Ecodevelopment . . . is a style of development which stresses specific solutions for the particular problems in each ecoregion, taking into account, ecological and cultural contexts as well as present and long-term needs".

According to Glaesar and Vyasalu (1984), "Ecodevelopment has two roots: ecology and development; so ecodevelopment should mean ecologically sound development". "Ecological sound" means achieving harmony instead of creating conflict between man and nature or more specifically between society and its physical environment. Harmony stands . . . for any kind of nondestructive relation within the limits of economic necessities and technological possibilitie . . . ecodevelopment is meant to be not only ecologically sound, but socially sound too".

Ignacy Sachs (1978) one of the foremost spokesman of ecodevelopment defines ecodevelopment as "an approach to development aimed at harmonizing social and economic objectives with ecologically sound management, in a spirit of solidarity with future generations based on the principle of self reliance, satisfaction of basic needs, a new symbiosis of man and earth; another kind of qualitative growth, not zero growth, not negative growth".

According to Sachs (1979) ecodevelopment requires:

- harmonization of consumption patterns
- appropriate technologies, ecologically based designs
- low energy profile, promotion of renewable energy base,
- new uses for environmental resources, careful husbanding of resources, recycling
- ecological principles to guide settlement patterns and land uses
- participatory planning and grass-root activation

The Dominant Elements of an Ecosystem at the Village and Farm Level

Incorporation of the environmental and sustainability considerations at the micro levels i.e., village and farm levels, is of great significance. The resource utilization trends at these levels, are influenced either by commercial over-exploitation of resources in some areas (e.g., depletion of CPR's) or by survival induced needs in others (depletion of forest cover). Further, such processes are seen to be operating on a regular basis incessantly, constantly and frequently. The fact that they do not cause any immediate and violent impact on the environment has tended to restrict the wider awareness of these issues.

The three basic resources which are of crucial importance, at village and farm level are land, water and biomass, all of which, are already under great stress in the developing countries. Thus a need is felt to introduce suitable strategy measures (i.e., sustainable intensification, conservation and restoration) at the village, community and farm levels. This is to be noted as a shift away from purely technical approaches towards community and farmer centered measures.

Environmental profile as well as planning should essentially concentrate on the natural resource base of the village settlement. Broadly, the four elements on which an environmental profile should focus are:

(i) Land;
(ii) Water;

(iii) Forests/Biological Diversity; and
(iv) Renewable energy sources.

Goals of Ecodevelopment

(a) Environmental regeneration;
(b) Forest protection and enhancement;
(c) Sustainable livelihood security; and
(d) Equity.

Environmental regeneration includes reclaiming degraded land and making it productive through a mix of measures, ensuring a perennial supply of potable water, water for cattle and for irrigation.

Forest protection and enhancement includes conserving forests, ensuring biological diversity in them. Planting trees in degraded areas and wastelands, growing fuel wood and fodder vegetation in community lands and involving the community in planning and management tasks.

Sustainable livelihood security has been defined in the following manner by the Brundtland Commission.

"Livelihood is maintaining adequate stocks and flows of food and cash to meet basic needs. Security refers to secure ownership of, or access to resources and assets to offset risk, ease shocks and meet contingencies. Sustainability refers to the maintenance or enhancement of resource productivity on a long-term basis. A household may be able to gain sustainable livelihood security in many ways: through ownership of land, livestock or trees, through rights to grazing, fishing, hunting or gathering; through stable employment with adequate remuneration or through varied repertoires of activities".

Equity indicates the degree to which the poor benefit from development in absolute terms. The need to reduce rural poverty and to bring about greater equity is one of the most pressing economic and social issues in developing countries. In the context of sustainable development, measures towards poverty alleviation like land redistribution and tenancy reforms, improved access of the poor to credit, irrigation, employment opportunities, knowledge on improved technologies, extension and training facilities, benefits from food distribution schemes, etc. will also go a long way in ameliorating environmental damage.

Ecodevelopment: Planning Steps at Village Level

Broadly any environmental planning process would consist of three major steps:

1. Analytical process
2. Plan Preparation Process
3. Implementation and Planning process

The various steps, methods, outputs and end results are indicated below.

Steps	Method	Output	End Result
1. Setting-up the baseline of parameters	Survey of state of environment critical information.	A quantified statement of the available natural reso-urces in the area.	Knowledge of potentials
2. Analysis of current use/ consumption	Quantitative and qualitative methods for assessment of the rate of utilisation of the resources	Identification of stresses	Identification of broad strategy
3. Plan Preparation	Prioritization Assessment of thresholds to development	Identification of problems (how to)	Objectives (Programme-Abstract)
4. Implementa-tion and plan modification process	Functional linkage matrix, identification of lags, gaps and misma-tches	Activity schedule	Impact measure-ment and plan modification
5. Monitoring	Identification of critical environmental indicators (in relation to data colle-cted at 1 above)	Observation of the trends of degradation	Replanning

For the village level, in particular (a) methodology is being putforth, to draw up village ecodevelopment plans in conjunction with specific strategies (i.e., ecosystems approach), goals (i.e., sustainable intensification for livelihood security, preservation and restoration) and project/programme formulations (compatible with the socio-cultural milieu). The major steps involved in the attainment of the above goals are shown in Fig. 8.1.

It is clear from the figure that a primary prerequisite for any effective plan formulation and management is the process of inventorying, which in turn reflects the quality and quantity of the information. The data requirement for a village farm ecosystem development plan would pertain to:

* Land Resource Inventory
* Nature and Extent of Land Degradation
* Land Classification and Optimal Land Use
* Biological Potential of the Land
* Physical, Chemical, Microbiological and Erodability Characteristics of Soil
* Water Quality and Availability
    — Impact of irrigation and crop intensification on the water resources of an area
    — Rate of exhaustion of aquifer; quality of groundwater; increase in salinity etc.
* Cropping practices (date, density and manner of sowing)
* Genetic erosion of Flora and Fauna
* Biomass (adequacy, availability; potentials).

The analysis of the above elements should be able to reveal areas of environmental stress and resources experiencing degradation. The areas that need attention (policy as well as research) to promote agricultural production on an economically and ecologically sustainable basis at a village level are:

* Land and water management
* Crop management
* Energy management
* Post harvest management
* Farming systems management
* Technology options

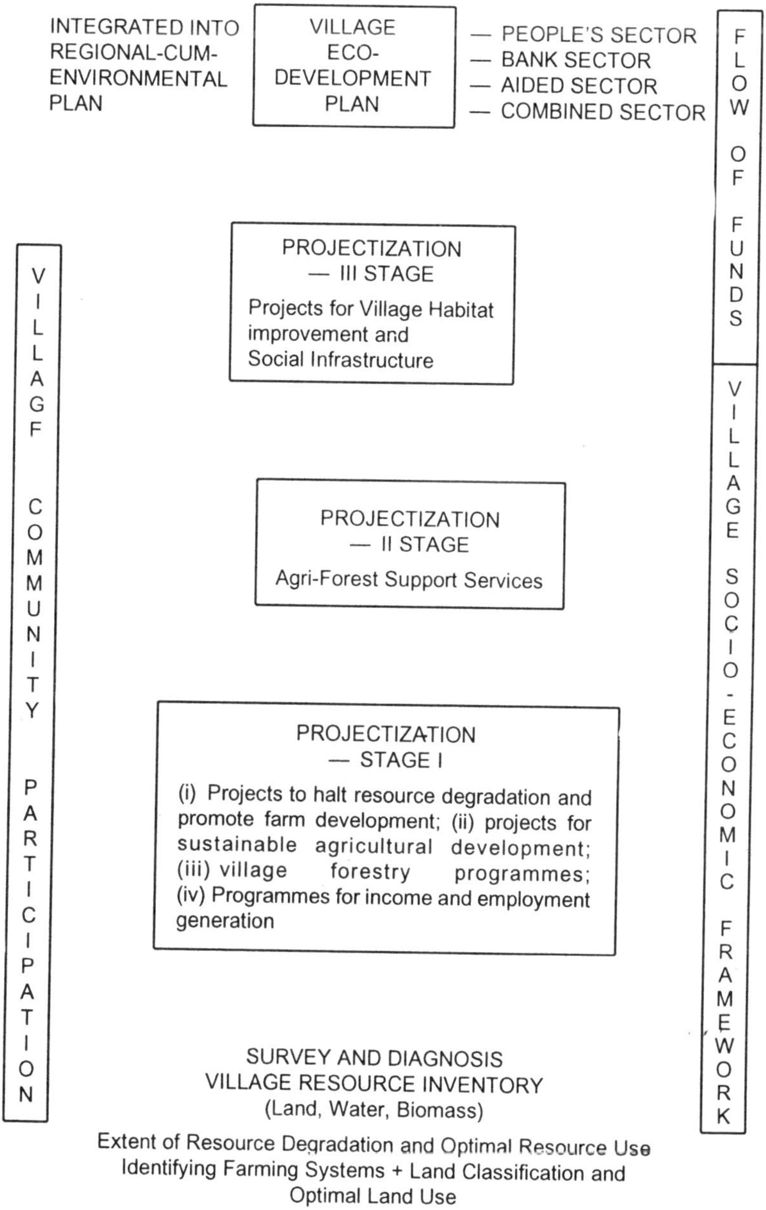

FIG. 8.1: Village/farm ecodevelopment planning—An approach

* Delivery systems
* Participation

Planning for sustainable development at village level, needs to address issues relating to the micro-environmental potentials, needs and priorities and further integrate these elements into the broader framework of Regional economic-cum-environmental planning. The local people, therefore, must have a chance to participate in decision making about the use of their resources and what their future will be like. The approach here should be as participatory as possible and environmental plans should be specific to the village socio-economic framework.

At the farm level, the Farming System Development (FSD) approach is seen to integrate the concept of ecosystems in farm development. A key feature of FSD is the dynamic nature of the procedure. Fig. 8.2 gives a summary of some of the procedures involved in FSD and some of the interactions that might take place between various steps involved in farm development.

One of the important aspects at the farm level is to understand and identify existing farming systems, and the opportunities associated with the farming systems in a particular area. The inventorying at this level would be addressing directly the problems confronting the farm, in particular, and the village ecosystem, in general.

To illustrate the methodology of ecodevelopment planning, some case studies in different environment typologies should be undertaken. Such case studies dealing with the village environment in its totality do not exist. This is a new area of planning which has to be explored. Studies on some individual aspects of environmental considerations, e.g., rural energy, exist. But examples of case studies viewing ecodevelopment as a total process, integrating economic, natural resource and environmental planning and responding to the changing needs of an area are rare to find. However, there is a unique experiment in Panchayat Resource Mapping, currently going on in Kerala (CESS, 1991) which has evoked immense interest among planners. This is a programme of scientific resource mapping and integration 'leading to a desirable and ecoconformable developmental strategy evolved through a series of action plans at panchayat level with people's participation'. The results of this experiment are keenly awaited.

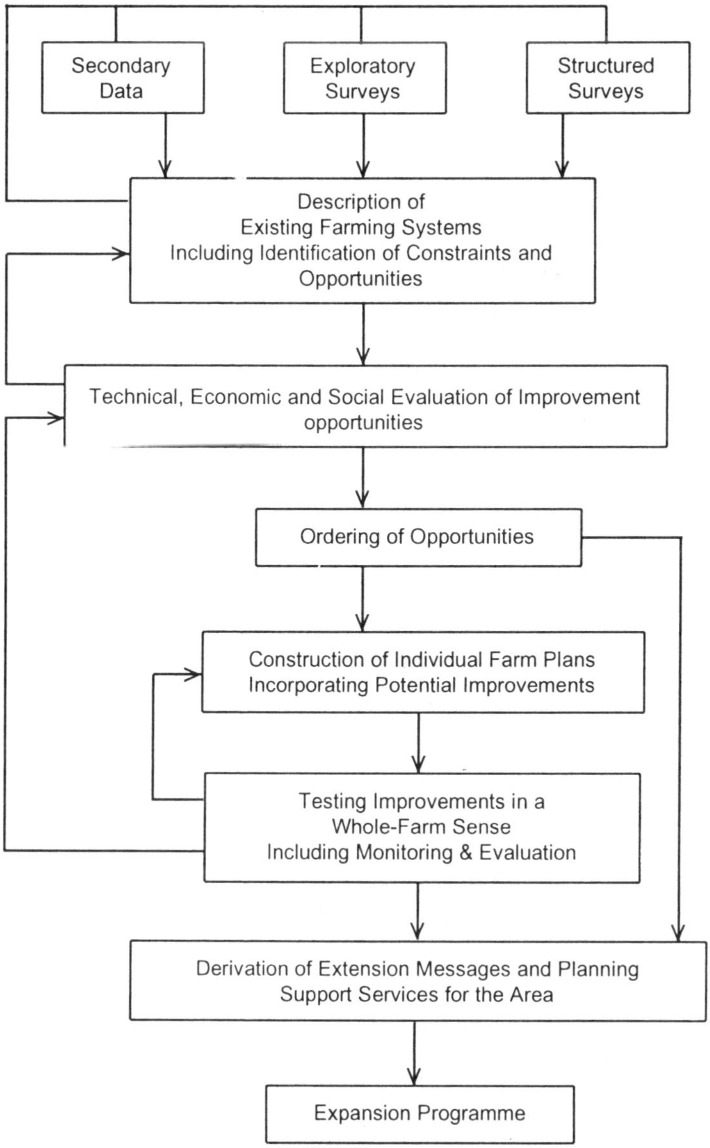

FIG. 8.2: FSD procedures and interactive nature of an FSD programme

Source: Regional Office for Asia and the Pacific (RAPA) Food and Agricultural Organisation of the United Nations, Bangkok, 1989.

REFERENCES

Centre for Earth Science Studies (CESS), 1991. Panchayat Level Resource Mapping—An Approach Paper, Trivandrum.

Gaesar, B. and Vyasalu, V., 1984. The obsolescence of Ecodevelopment. In Beruch and Glaeser ed. *Ecodevelopment: Concepts, Projects, Strategies*, Pergumon Press, Oxford.

Sachs, I., 1987. The salient features of development. In *Environment and Development—Phase III,* Report of a Workshop CIDA & Dept Envt, Ottawa, 1978.

UNEP, 1974. Reprint, UNEP Governing Council.

9

RESOURCE EVALUATION FOR SUSTAINABLE RURAL DEVELOPMENT

ANJANA DESAI

Philosophy of Sustainability

THE philosophy and concept of "Sustainable Development" are based on social justice. It is the Human Right of individuals to achieve self-sufficiency in food and employment. The concept does not aim at maximizing production, but it has the goal of achieving proper distribution of production and consumption of food through proper management of all regional resources to attain self-sufficiency. The concept of sustainability is based on also ethical and moral values of conservation of resources, so that the present generation can share them with that of the future as they have done so with the generation of the past. In today's world, development is considered not by technological skill only but by advancement in conservation of resources for sustainability and self-sufficiency.

Historical Background of Indian Rural Sustainability

In the Preindustrial Period, the pressure of population on land was low, and so was the total production of food. People had sustenance on local resources as there was not much trade due to lack of connectivity. The Industrial Revolution brought changes in the occupational structure, system of production, distribution of population and regional sustainability through trade and exchange of products. On the Indian scene, the process of development was different. Port oriented large metropolitan cities grew up on activities related to commerce and services without the base of industrialization. In rural areas, the pressure of population increased in relation to growth of population. Moreover, due to lack of enough employment and farming opportunities, selective migration of young male workers to large cities takes place. As a result, many villages in India, having seventy per cent of total population, cannot sustain themselves today with food, employment and basic amenities. However, the Post-Independence Period has experienced a tremendous growth in food production in India with Green Revolution. In spite of that, nearly 50 per cent of population in India is under the poverty line with severe economic stress in the midst of plenty. Hence, the development is lop-sided and skewed in distribution. The concentration of development is centered around a few areas and a few people have the purchasing power to consume most of the production. Sometimes food is also exported at a high price instead of being distributed within the country due to lack of purchasing power. Hence, high production does not guarantee proper distribution of food and sustainability of rural areas.

Objective: Need for Reviving Sustainable Rural Development

This chapter brings out the need of reviving the original system of rural sustainability through proper management of resources—both physical and human type. It is essential to evaluate resources in each region to find out whether (a) the major landuse pattern is matching with carrying capacity of land; (b) whether present land utilization of each landuse zone is matching with its potential level of output in relation to its ecosystem; (c) whether the process of land utilization has improved or degraded physical resources; (d) whether there has been distinct improvement in backward regions in terms of health, education and employment of people; and finally (e) micro-level planning has to be evolved, based on local ecosystem, on one hand, and socio-economic background of the people, on the other, with appropriate, implementable technology, organized by the people themselves in each village.

The relevance and justification of focussing on backward regions and people are based on the concept of "sustainability", leading to conservation of resources for self-sufficiency. As every drop makes an ocean, each piece of land can certainly sustain its people provided it is managed judiciously. Sustainability of rural base is the only way to decentralize the process of development of rural areas from urban pivots. Much of rural areas are not fully utilized or managed properly. Land degradation has taken place leading to lower yield per acre. Moreover, the increasing pressure of population has added further problems to it. Instead of checking land degradation, rural people have encroached upon areas which are not suitable for agriculture. In many places pasture land and hill-slopes have been cleared for cultivation. Initially, this has led to an increase in production for a short while, but such practices have created wastelands over a period of time. Eventually in such regions net sown area and gross cropped area have reduced leading to loss of sustainability. Basic resource of land has not been conserved and maintained. Only three months of agricultural work in a year is not sufficient for food or employment. People have to migrate to other areas as a means of survival strategy.

Intensive agriculture may not be possible in each region due to physical hazards. A combination of mixed farming of agriculture while dairy or poultry farming, fishing, mushroom farming or sericulture, forestry, agrobased small scale industries, depending on local ecosystem and technology available at hand, will certainly increase rural employment and purchasing power of villagers. Such an attempt will achieve rural sustainability and will be able to hold rural working force to its root, leading to decentralization of development. Redistribution of population can be achieved as rural-urban migration will be reduced on one hand, and exploitation of unskilled, rural migrants by labour contractors will be minimized, on the other. Sustainable rural development with basic food, employment and amenities will be able to bridge the socio-economic gap between "people" and "places" with security and self respect. Such a process of development will uplift the rural base by combining regional self-sufficiency on one hand, and regional specialization, on the other hand. Proper management of resources will be able to have a system which is economically viable, socially stable and environmentally sound.

Methodological Framework

Evaluation of resources has to be carried out at micro-level (village wise). This grass-root approach is fundamental, if sustainability is to be achieved in rural base. The basic resource in rural area is land and water. Conservation of soil and water is to be organized by the local people for their sustenance. In this respect agroclimatic regions, based on the quality and quantity of soil and water, can be considered as the ecological boundary of an area. Within the framework of an ecosystem, an evaluation of resources should be done in terms of man and environmental interaction. In other words, it is necessary to evaluate the actual output (in Rs.) per unit area compared to potential output of an area. In doing so, it is important to keep track of environmental management, as stated in the objectives of resource evaluation. It is necessary to monitor whether a change in landuse or in cropping pattern has resulted in land degradation which can be caused by physical processes like drought, flood or by human interference, eg., deforestation, encroachment of agricultural activities in areas which are suitable for other activities, excessive use of irrigation, chemical fertilizer and pesticides. Similarly land which is good for agricultural purposes should not be used for residential or recreational purposes. Due to increasing urbanization and industrialization often fertile agricultural land is lost.

An evaluation of resource and developmental programme can be carried out in the following framework. Firstly, there is a need of a detailed survey of physical and human resources at village level within an ecological region. Major variables should be mapped for a spatial analysis. Traditional cartography takes time in mapping data on a temporal basis. Hence, recently Geographic Information System (GIS), a powerful software package, is used for spatial and temporal analysis for the purpose of planning. The second step is to have a broad classification of land resources, mainly based on physical variables like relief, slope, climatic variables, soil fertility in terms of NPK, pH, texture, erosion, vegetative cover, quality and quantity of surface and underground water. GIS can be used here also to have a classification of land, based on suitability to agricultural use. Thirdly, an evaluation of land resources of regions can be done by matching the current utilization of land with that of the land capability as given by land classification. It can be easily monitored, if there is a "mismatch" between the landuse and land classification. The mismatched areas can be further correlated with areas which are suffering from land degradation. The reasons of land degradation can be probed and rectified at village level. In implementing microlevel planning rural technology has to be evolved by the people for conservation of soil and water, on one hand and appropriate development of non-farm activities to achieve full employment, on the other hand. Finally, present output per unit area and farm income per person should be compared with those of potential figures. It is useful to have information on whether "effective" land can be increased for food production through intensive agriculture or through a change in cropping pattern or a new way of water harvesting and soil conservation. Farmers need to organize themselves to increase their income not just through agriculture but also through village based nonprimary activities of various types as per the skill of the people and suitability of the ecosystem.

Development of data-base over time and space at a microlevel is the basis of a successful regional analysis and planning. The GIS software package has the capability of handling a large number of variables on a spatial basis. Weightage can be given to each variable which can be superimposed to demarcate specific zones as per the need of the research, for example, to demarcate agroclimatic regions, to know whether there is a match or mismatch of landuse and land classification, to locate land degradation or to assess potentiality of land.

Secondary, published data and remotely sensed data can also be useful for an overall spatial pattern to depict changes of landuse over time. However, in order to have a meaningful explanation of spatial pattern one should have field experience to know the behavioural aspects of villagers—the way they use or misuse their environment, the way they perceive and cope up with physical hazards, whether they have any concern of environmental management and improving the life style in villages with enough food and employment.

Micro-Level Planning

Poverty is deeply rooted in rural areas of India. If developmental process does not start from the rural base, poverty cannot be eradicated and rural areas will be deprived of socio-economic progress and eventually the gap between rural and urban areas will increase. Microlevel planning has various facets which need to be coordinated and implemented: (a) positive promotion of environmental change, (b) coordination of different government and nongovernment agencies, (c) participation of villagers in all programmes, and (d) management of physical and human resources. It is important for villagers to know how nature works and how one should cope up with its hazards. For example, there will be an ever increasing need for food, fuel and fodder. If deforestation is done faster than replenishment of trees or more food is produced by clearing pasture and forest areas as an act of encroachment of landuse, there may be a short term benefit, but with a permanent, long term damage to the environment. Hence any development which depletes and impoverishes resources in an environment, should not be considered as a developmental process. An environmental change has to be a positive one. It is noticed that agricultural practices or choice of crops often lead to land degradation. For example, in order to have a quick economic gain, farmers change over to a farming of monoculture of cash crop with an excessive use of irrigation, chemical fertilizers and pesticides. This has led to the loss of bio-diversity on one hand and degradation of land through soil salinity, pollution of soil and water through chemical pesticides, on the other hand. Degradation of land also takes place in unirrigated areas through human interference like over grazing or encroachment of agricultural activities in nonagricultural areas. Right decisions have to be taken up by farmers in order to check human misuse of land. Incentives can be given by the Government, through subsidy or cheap loan, if farmers upgrade and protect their environment.

Coordination of various government and nongovernmental agencies is very much necessary in order to get the rural programmes to be implemented. Research done on land degradation, land classification through Geographic Information System, land-suitability of various uses and cropping pattern at Agricultural Universities should be coordinated through Government and non-government agencies so as to reach village "Panchayat" and "Gram Sevak" who will be able to organize villagers with proper information and techniques to be used.

Microlevel planning through proper management of human and physical resources has shown amazing results in a few pockets in India. Such efforts need to be implemented in all villages for their sustainability. The basic physical resources like land, water and vegetation can be easily maintained and protected individually as well as through community approach. Water harvesting and conservation of soil can be done with the help of water collecting wells, check dams, plantation of trees, drip-irrigation and appropriate use of land. Use of nonconventional energy like bio-gas, wind and solar energy can certainly raise the standard of living in villages without creating any problem to environmental resources. In this regard, development of human resources is extremely essential to train the local people so that they will be able to harness the local existing resources and will be able to find out alternative ways of using resources. For example, more and more farmers in Gujarat are turning towards dairy farming in drought prone areas because of uncertainty of rain which makes it risky for a good agricultural output. Similarly coastal areas may not be suitable for agriculture due to high salinity of soil and water. Villagers can take up fishing activities in such areas as an alternative occupation. It is very important to organize alternative occupation in each village along with agriculture so that people can have work throughout the year to sustain themselves and there will be no need for them to migrate, leaving their home and family for survival. A combination of farm and nonfarm activities will create employment opportunities and thereby increase the purchasing power of farmers. Management of resources should include not only environmental conservation but also production of items with proper design and quality so as to capture market with proper price. Marketing of products requires a combined effort of all farmers in the form of a cooperative system. Self-help and individual training along with participation of all communities can bring prosperity to rural people.

Empirical evidence of this chapter is based on three backward regions of Gujarat—the Panchal, the Nalkantha and the Bhal. All these regions suffer from physical hazards. The Panchal region is hilly and drought prone along with soil erosion. The Nalkantha suffers from drought and soil salinity, while the Bhal region has problems of both flood and drought along with soil salinity due to marine ingression. These are fragile ecosystems, lying in the area which was a part of the sea, separating Saurashtra and the mainland of Gujarat. These are in the category of backward regions where agricultural yield is low, on one hand and purchasing power of people is also low, on the other hand, due to lack of work except for three months in the rainy season. People have to migrate as an unorganized, unskilled labour for the sake of survival. It is found that, land degradation is progressing at a fast rate during 1967-1988 due to several reasons. The pressure of population has increased, and land and man ratio has decreased resulting in less than one acre per person. Hence, people have to encroach on other land uses for agricultural use. Eventually there has been a decrease in net sown area and gross cropped area, leading to lower level of nutrition per acre, farm earnings per acre and nutrition-man ratio. Hence, rainfed agriculture, small piece of cultivable land, large size of families, degradation of land are some of the reasons which are responsible for nonsustainability of people with proper food and employment. The vicious circle of land degradation and migration of young males seems to be a permanent problem of poverty in rural India. Employment has to be generated in rural areas; migration needs to be checked and environment has to be conserved for sustainability of villages.

REFERENCES

Ali, M., 1978. *Situation in Agriculture, Food and Nutrition in Rural India*, Concept Publishing Co., New Delhi.

Davidson, Donald A., 1992. *The Evaluation of Land Resources*, Longman Scientific and Technical Essays, England.

Mitchell, D., 1979. *Geography and Resource Analysis*, Longman Scientific and Technical Essays, John Willey, New York.

Nair, N.J.K., *et al.*, 1985. Environmental Mapping—An Aid for Long-Term Planning, Resource Analysis Division, Centre for Earth Science Studies, Trivandrum, India.

Pacione, Michael, 1983. *Progress in Rural Geography*, Croom Helm, London.

Rammaiah, Y.V. and Reddy, N.B.K., 1983. Carrying Capacity of Land in Andhra Pradesh, *Indian Geographical Journal*, Madras, Vol. 58, p. 107-18.

Rao, U.R., 1992. *Remote Sensing for Sustainable Development—Indian Pespective*, Lead Article in Book: 'Natural Resources Management—A New Perspective', ISRO Publication, Bangalore, R.L. Karla (Ed), 1992, pp. 1-6.

Swaminathan, M., 1980. *Principles of Nutrition and Dietetics*, Bangalore Printing and Publishing Co. Ltd., Bangalore.

US Department of Agriculture, Soil Conservation Service, 1961. Washington D.C.

Sustainable Landuse Systems Research, Proceedings of an International Workshop, New Delhi, Feb. 12-16, 1990, Oxford and IBH Publishing Co., New Delhi.

10

COASTAL FISHERIES
Towards Integrated Community Management

JOHN KURIEN

Introduction

IN the recent past there have been a variety of initiatives aimed at creating an awareness about the *sine qua non* of crafting institutions at the community level for sustainable development and management of natural resources. The call for community development and management of coastal fisheries is but a specific case of this rapidly growing global preoccupation.

In this chapter, which is based on the author's experience both as a social activist working with coastal fishing communities and also as a researcher, the stress will be to show that integrated community management of coastal fisheries can be assured only by (i) a rebuilding of the "community" in coastal villages; and (ii) a simultaneous and conscious multi-sectoral and transnational awareness and action on the ecological and economic fronts.

The primary reason for this is that the "community" which we refer to has undergone a significant socio-economic and cultural metamorphosis as a result of "incorporation" into larger national and international ecological and economic processes. We are therefore talking about a "community" which has significantly altered its relationships both within and without. It also continues to be in a process of dynamic transition.

In this context, in order to define the contours of a meaningful community management of coastal resources, we need (i) a brief understanding of what the early community was like; (ii) a history of the incorporation process and the consequent demise of community in coastal fisheries; (iii) an understanding of the efforts to revive small-scale coastal fisheries; (iv) an articulation of the measures for rebuilding the community; and (v) a spelling out of the basis for integrated community management. The chapter is organised along these lines.

The Early Community

The Asia-Pacific region has had a long history of maritime contacts for trade. It is the sea—the *samudra*—which brought the societies of yesteryear in the region closer together. It has been argued elsewhere by this author that such a maritime tradition must have had its origin in the humbler occupation of coastal fishing with the generalised skills of navigation and the specialised knowledge of the various aspects of the aquatic eco-system as the points of departure. These maritime contacts have left their mark on many coastal fishing communities. The most obvious of these being the strong element of the exotic in the craft and gear designs of many artisanal fishing communities in the region.

It can be argued that coastal communities of the region had earlier been more open to influences from other societies across the sea and relatively less to their own land-based neighbouring communities. This latter phenomenon is more pronounced in large non-island societies like India.

This isolation from mainstream society and the compulsions of a maritime occupation—which took one out to sea and then back only to the physical fringes of the land—created a special psyche and culture among these communities so that their integration into the larger socio-cultural and eco-political mainstream of land-based society in their respective countries was much slower. There was perhaps a time in Asian societies when it was possible to conceive of small communities which were autonomously in control of the ownership and use of coastal fishery resources through a variety of institutional forms. They were probably not all very egalitarian, but yet, it is certain that due to socially sanctioned community caring and sharing mechanisms, no one was pauperised or destitute.

The Asian reality of today is a long distance away from these idyllic village societies. However there are "islands" where the possibility of reviving some of these communitarian dimensions are still a live possibility.

Incorporation and the Demise of Community

Initially the process of incorporation was largely through the exchange of the outputs of the sector—the fish. Fish is a perishable commodity and fishermen could not live by fish alone. A fisherman who netted more than two fish had a "surplus". The compulsion for exchange thus came very early in the development of the productive forces.

With the advent of independence and the modern state, in most developing Asian countries, the coastal fisheries was considered "backward" and sought to be "modernised" through a variety of means. As a result, the incorporation process was hastened through the input markets. The most popular of these were the introduction of mechanised crafts and the use of nylon for fabrication of nets. The aim of the infusion of this big dose of capital into the traditional artisanal fishery was presumably to raise the "production of fish for food by exploiting deeper waters more efficiently." Much of this modern "hardware" was state sponsored and came largely from countries such as Germany, Norway, Canada and Japan, to name a few. Rarely, if ever, were these inputs accompanied by the "software" of management measures and institutions which existed in these countries.

In general these new technologies were introduced without significant recognition of the existing plethora of artisanal fishing technologies and traditional community institutions which held the key to the customary use rights of the coastal commons. This new production-oriented approach to fisheries development introduced a strong element of multi-sectoral linkages between the fishery and the other sectors of the national and transnational economies.

It is important to note that the intended objectives of the new development programmes went adrift in most Asian countries by around the 1960s. One common factor can be isolated for this: the strong market compulsions which resulted in an incessant pursuit of prawns for export to the developed countries.

This export orientation and the consequent fuller integration into the transnational economic order was achieved in two ways: (i) by the introduction of "efficient" harvesting technologies from temperate waters like bottom trawling using mechanically propelled boats—to the total neglect of many artisanal technologies evolved over the centuries for catching prawns; and (ii) by the conversion of the coastal waters from a "community heritage" to an "open access" realm for investors from outside fishing communities best able to control and manage these new capital-intensive technologies. Consequently the pace and the direction of coastal fisheries in developing Asia was thereafter largely determined by investors from outside the coastal fishing communities and the demand preferences of the developed world.

The process of incorporation through the market for fish gathered momentum. Fishing villages in isolated parts of South and South-East Asia were instantly "linked" to Japan and the USA the moment they netted prawns or cuttle fish even if they continued to use the most rudimentary craft and gear. By the mid-1970s, and well into the 1980s, the coastal waters of Asia were ridden with physical and socio-ecological conflicts between the vast majority of fishermen from coastal communities who continued to use artisanal technologies and the operators of larger mechanised boats—particularly trawlers. An important dimension of this conflict, not usually highlighted, is the fact that its intensity and its ramifications depended largely on who were the owners and operators of the larger vessels.

Across Asia the owners of these vessels were rarely active fishermen from coastal fishing communities themselves. They were generally fish merchants, investors from outside the fisheries sector or non-fishermen from among the coastal village communities. The owners were largely nationals of the respective countries but often from an ethnic or social group different from the coastal communities. The crew of the mechanised vessels were from diverse socio-occupational groups in the economy but often from the coastal communities themselves. The most common crew composition was a mix of skilled workers from artisanal fishing communities and the relatively less skilled from among the poor peasantry and/or indigenous peoples of different ethnic backgrounds.

Consequently, the oft mentioned "conflicts between the coastal small-scale fishermen and the trawlers", were not without contradictions when viewed from the perspective of the coastal communities. Added to this the inequalities in wealth and income between artisanal fishworkers and the owners and operators of trawlers exacerbated social tensions on land. In most countries, the state had been inclined to view these conflicts at sea and on land as "law and order issues" and rarely as an expression of dissent against a distorted fisheries development process which deprived the coastal artisanal fishing communities of their livelihood and resulted in a process of deprivation not hitherto experienced by them. In the ultimate analysis the conflicts were also a reflection of the breakdown of "community" in the coastal fishing villages.

Returning to Small-Scale Fisheries

In the late 1970s and the early 1980s—that is, in the middle of the phase of conflicts in the fish economies of the region—we see a revival of interest in coastal small-scale fisheries. The most concerted effort on the academic front has emanated from the International Centre for Living Aquatic Resources Management (ICLARM). The FAO/UN which had earlier given technical assistance for the modernisation programmes in Asia also took interest in the revival of small-scale coastal fisheries. International funding agencies began to review their assistance programmes and gave a greater thrust to funding which explicitly promoted small-scale fisheries. The FAO World Conference on Fisheries Management and Development,held in Rome in 1984, though it was preoccupied with the implications of the UN Conference on the Law of the Sea (UNCLOS) was also pressured from within and without to make significant pronouncements about the status of coastal communities and the role of small-scale fisheries in the future strategy for global fisheries management and development. An International Conference of Fish Workers and their Supporters (ICFWS) held parallel to the FAO meet of 1984 highlighted the plight of coastal communities and stressed the need to reorient development and management to focus on their needs and aspirations.

A closer look at these concerns show that with a few outstanding exceptions, the emphasis was entirely on the issue of scale alone. Concretely it merely meant stressing beach landing craft rather than harbour based vessels or outboard motor-fitted canoes instead of large mechanised trawlers. There was certainly no appreciation of the linkages between smallness of scale, the nature of tropical coastal resources, the knowledge systems of artisanal fishing communities and their traditional socio-cultural institutions.

In other words no pretensions of reviving community!

Fortunately for us, some of the encyclopedic knowledge of artisanal fishing communities and the traditional community institutions among them for managing their resources as well as caring for individuals, have survived through the modernisation phase. Quite often this had happened because these communities were not fully "incorporated" into the fissiparous dimensions of modernisation, or, as in the case of some, they were able to "contain" modernisation within the purview of their earlier institutions. There is now a considerable body of literature on this. We need to blend this concern for scale with the revival of community knowledge and community institutions to make a beginning at rebuilding community.

Rebuilding Community

Given the socio-economic realities of Asia, particularly South Asia, where pressure on coastal resources is still on the increase and gainful employment opportunities outside the fishery sector are hard to come by, it is one's assessment that for another generation at least, the solution to the problems in the coastal fisheries must be found largely within the sector itself.

In the live reality which confronts them, fisherfolk have devised many individual survival strategies to keep their necks above water. These have taken a variety of forms ranging from investing in more fishing assets; devising various crew rotation and mutual assistance measures; taking to destructive fishing; migrating to other countries in search of jobs; turning to other sectors of the economy in order to eke out a livelihood; and making efforts to rejuvenate the resource. Equally important have been their efforts at collective action to pressurise the state to intervene and bring order into their commons. The best examples of this come from Indonesia, Philippines and India.

There are potentials and limits to both these forms of action by individuals and the state. The evidence, again from India and Philippines, seems to be that unless there is some form of mediation between individual and state action, coastal fishing communities face a bleak future. The need of the time is therefore basically to redefine and rediscover the "community" in coastal fishing villages in the context of the socio-economic and political realities of today.

The fisheries "modernisation" process neglected—or worse, threw overboard—a rich community tradition of concern for the labours of Nature, the concept of the *samudra* as a heritage of the "dead, the living and those yet to be born" and the socio-cultural institutions of sharing and caring within the community. Also, the social and economic inequalities that resulted as a consequence of the modernisation process, which we mentioned above, cannot be wished away in this rebuilding.

It is very unlikely that all those who live along the coastal zone today have common interests as regards the coastal fishery and its future. There is, therefore, need to identify and bolster a core group among these communities which can become the "beacons and guardians of the sea" around whom one can envisage concentric circles of supportive interests. There are at least three supportive conditions which must be ensured to achieve this objective. We shall sketch them out briefly.

Aquarian Reform

A key condition for rebuilding community will be successful implementation of an aquarian reform package. There are initially four facets to this reform. First, the right of access to harvesting technology should be restricted exclusively to those who fish. Secondly, there must be a ceiling on the number of units such a person can own. Thirdly, with an increase in the scale and capital investment in the harvesting units of such individuals and groups, their realm of fishing should be appropriately restricted to the specified zones of access in the Exclusive Economic Zones of their respective countries. Fourthly, such persons should be vested with the right of first sale of the fish they harvest.

Under the stewardship of such a class of workers and owner-workers, the coastal waters can regain their status of "community property under community custodianship". It is needless to say that such reforms are rarely granted on a sliver platter. They will only materialise if there is sufficient pressure from the grassroots. Again the evidence from Philippines and from India give room for optimism in this matter.

Community Development

Rebuilding community implies not merely regaining the rights to the coastal commons but also simultaneous action on the socio-economic development front. In many developing Asian countries despite three to four decades of conscious planned development efforts, coastal fishing communities continue to be at the bottom of both the social and economic ladders of their respective countries. It is imperative to change this situation if the gains from aquarian reforms are to be consolidated and sustained into the future.

Spelling out a general strategy to bridge the socio-economic gap between fishing and other communities would be a futile exercise given the diverse socio-political considerations that determine the degrees of freedom available for action on these fronts. Concretely this calls for better organisation of the fish marketing activities to be able to earn more income; better credit arrangements; more access to better education and health facilities; more job-oriented skills for the youth and women.

Based on personal experiences in community development work and also an understanding of what is happening in some other countries, it can be said with confidence that people-managed and people-controlled organisations which can be supported by voluntary agencies are the best bet to achieve a participatory and holistic development process in coastal fishing communities. This is not to minimise the role of the state and enlightened public policy in rebuilding community. To achieve this calls for a reorientation of fishery development policy to focus more on the nature, human and social capital in the sector and less on the human-made capital in the form of artifacts alone.

New Relationships

In the ultimate analysis rebuilding community is not achieved by redefining and changing property relations and socio-economic institutions alone. Reinstating the relationships of nurturing, caring and sharing between humans and nature and between human beings is an intrinsic goal of this endeavour. In this context there is a very central role for women in the coastal communities—particularly those from fishing households. Motherhood bestows on women the very special trait of caring without counting the cost or being calculative about future returns. Such values are essential to underscore the need to move towards a "nurture, caring and sharing approach" to coastal fisheries.

Integrated Community Management

Integrated community management cannot be seen as an isolated project of the inhabitants in a series of coastal villages who have been able to rebuild a community as sketched out above.

Ecologically, integrated community management of coastal fisheries starts in the forests. Economically it begins in the village but its linkages are often in Japan, United States or Europe. There is therefore a strong multi-sectoral and transnational dimension to it today which is totally outside the control of this community. This fact seems often ignored in much of the discussions on community management.

This needs to be explained more concretely.

Multi-Sectoral Ecological Action

The coastal fisheries of a country is not just affected by the fishery activity in the seaward side of the coastal zone. It is also very greatly affected by all the negative externalities of the economic activity that takes place in the hinterland, largely via the links in the aquatic systems.

Increased silt loads due to deforestation; reduced river water flow following dam construction; effluent from agriculture, industry, urban settlements and tourism which are dumped into the rivers and the sea, all go to effect the ecological integrity of the coastal ecosystems. In the initial stages, such externalities cause no harm. They are quite easily "managed" within the capacities of the natural physio-chemical processes in the coastal waters.

However, a point is reached when these negative externalities "accumulate" at a rate faster than they can be "absorbed", thus pushing the system to an ecological precipice. Most coastal ecosystems in Asia seem to have reached the fringes of this precipice. To retrace steps back to the plateau of ecological stability requires an awareness on the part of coastal communities that collective action on their part alone for the management of their common resources is only a partial solution—and a poor one at that.

Since the "sea starts in the forests", cooperation from the "up-stream" sectors of the economy is a prerequisite for total success of their venture. This in turn implies a willingness on the part of the actors of these up-stream sectors to regulate their polluting and resource degrading activities. Such are the multi-sectoral compulsions of integrated community management of coastal fisheries.

A beginning on this front for multi-sectoral environmental action to ensure integrated community management of coastal fisheries calls for both state intervention and community action. The author's personal experiences of the latter has been in assisting the National Fishworkers Forum of India—a confederation of artisanal fishworkers unions/associations—in organising an all-India people's awareness *jatha* (march) in 1989 on the theme "Protect Water, Protect Life". The uniqueness of this socio-ecological march which covered about 5000 kms of coastline was that it was spearheaded by the labouring men and the women of the coastal fishing communities—the beacons and guardians of the sea. They were joined by many middle-class environmental groups and other workers' groups and people's organisations from the "up-stream" sectors of society. The march provided a broad platform for creation of awareness and knowledge of the fact that our common linkages to life are through water!

Transnational Economic Solidarity

There is also a transnational dimension of integrated community management of coastal fisheries which requires greater attention than it presently receives. This has to do with the fact that a significant share of the value of output of the fishery resources in the coastal fisheries in developing Asian countries is obtained from exports to the developed world. There is no reason to expect any change in these economic linkages in the near future.

Consequently, the nature and the quantum of fishing effort in the coastal fisheries, even assuming management by the community, will largely be dictated by consumer demand in Japan, U.S.A and Europe.

Will localised community efforts for a sustainable coastal fishery succumb to the insatiable demand for exotic tropical coastal seafood in the developed world? The answer is a big "YES".

The solution to this dilemma is not to give up efforts for localised community management but rather to integrate this with global initiatives to reduce the degree of discord in the nexus between transnational trade and consumption patterns on the one hand and sustainable local community management of coastal waters on the other. To achieve this we need NOT ban all fish exports from developing Asian countries or stop the imports of fishery technology from the developed world. That would jeopardise the livelihoods of millions of fishworkers. The need of the hour is to arrive as a proper balance—be it in harvesting of the resource or the use of technology for fishing and related activities.

For example, one of the most crucial transitions in Asian coastal waters will be to make a shift from using trawls (to catch prawns) to getting back once again to the more passive and environmentally benign nets for this purpose. The economic and ecological effects of "trawl-free prawn exports" has been partially demonstrated in Indonesia and one can do this in other countries as well. Achieving such transitions are an important prerequisite for a sustainable and integrated community management of coastal fisheries.

But such transitions cannot materialise without the right signals to this effect from the developed world. For example, if the Japanese consumer movement can get Japanese prawn importers to refuse prawns below a specified count/kilo from India, the salutary environmental and economic impact on coastal fisheries will be far greater than all the costly efforts we now make to manage the extremely socially and politically charged conflict between trawler owners and small-scale fishermen.

Conclusion

If integrated community management of coastal fisheries at the local level has to succeed we need an integration of multi-sectoral and trans-national awareness and action.

This can be achieved only in the process of the struggle to rebuild community in the coastal villages. The community focus and the multi-sectoral and transnational action are not necessarily in conflict. In fact we should strive towards the ability to blend the two without losing the cutting-edge of any of their respective concerns. Let this not be mistaken as a search for harmony and co-existence. Far from it, it is only a creative effort to balance countervailing forces. In that sense, we may never reach an integrated community management of coastal fisheries—we will always be moving towards it. But first we need to make a firm commitment in that direction.

Acknowledgements

This is a revised draft of a paper presented at an international workshop on "Community Management and Common Property of Coastal Fisheries in Asia and the Pacific: Concepts, Methods and Experiences" held in Manila, Philippines in June 1993.

REFERENCES

Clark, J.R., 1992. Integrated Management of Coastal Zones, FAO Fisheries Technical Paper 327, FAO, Rome.

Emmerson, D.K., 1980. Rethinking Artisanal Fisheries Development: Western Concepts, Asian Experiences, World Bank Staff Papers, Washington D.C.

Kurien, J., 1988. Studies on the Role of Fishermen's Organisations in Fisheries Management (with particular reference to the Indo-Pacific Region), FAO Fisheries Technical Paper 300, FAO, Rome.

Kurien, J., 1990. Knowledge Systems and Fishery Resource Decline: A Historical Perspective In Lenz., W. and Deacon, M. (eds) *Ocean Sciences: Their History and Relation to Man*, DHZ, Hamburg.

Ostrom, E., 1990. *Governing the Commons: The Evolution of Institutions for Collective Action*, Cambridge University Press, Cambridge.

Pomeroy, R.S., 1993. *A Research Framework for Coastal Fisheries: Co-Management Institutions In Naga*, ICLARM Quarterly, Jan 93, Manila.

Ruddle, K., 1985. The Traditional Knowledge and Management of Coastal Systems in Asia and the Pacific, UNESCO-ROSTSEA, Paris.

PART II

CASE STUDIES AND EXPERIENCES

11

TRADITIONAL KNOWLEDGE FOR SUSTAINABLE DEVELOPMENT
Kerala Case

T. MADHAV MENON

Introduction

THE concept of "Sustainable Development" is still imprecise. One popular definition is that "it meets the needs of the present without compromising the ability of future generations to meet their needs" [Agarwal, (ed.) 1992]. Sukhamoy Chakravarty points out the concept is not static, but tries to indicate "certain directional changes in economic policy making, nationally as well as internationally. . . ." (in *ibid.*). It has to be dynamically integrated into any plan of economic development, both at macro as well as micro levels. It has implications in all dimensions of human progress. Sociologically, it comprises the question of differential advantages to various sections of the population. Technologically, it implies a choice among many available alternatives, such that the application of the chosen technical means is such that it does not entail the exclusion of alternatives over time. Economically, it forces a choice among competing non-economic values, inhibiting the maximization of specific exclusively monetary gains over limited time spans. Thus it encompasses not merely the allocation of a specific resource, *e.g.*, land, for a specific exploitative use, but also a total value system, to be intelligently chosen by a community even at the sacrifice of immediate material advantages. In this article, the emphasis will be on land and forest, and the tenure of human communities in such environments.

The Changing Objectives of Agricultural Development. For nearly nine-tenths of the time Homo Sapiens has been on the earth, the species was subsisting on sporadic cultivation in forest habitats, which the humans shared with other animals. "Shifting cultivation" was the first stage of transition from nomadism to farming; it is believed to have originated in South East Asia, some sites in New Guinea having been dated back to 3000 BC. It was a worldwide phenomenon, particularly in the tropical regions, in Asia, Australia and the americas. In Indonesia and among the Kanikkar of Kerala, it extended wet crops like paddy, but is generally applied to dry crops like millets, roots and tubers, vegetables, grams and pulses, generally the so-called subsistence or survival crops. The sites were determined both by the dynamics of the wanderings of the community over a territory as well as the location where it preferred to stay for longer and longer periods until they "Settled" there.

Methods varied; tools were devised accordingly. The Kurumba and Muduga living in the wet evergreen forests of the Western Ghats in Kerala required only a digging stick to scrape enough area required for the desired plant species, which they planted, all together to minimize effort. The rest of the native vegetation was left in tact; no slashing and burning was practiced. The desired plants grew with all the rest of the vegetation; this had the advantage of evolving cultivars which has the maximum survivability in the location. There was no "weed problem" because the communities found a use for every plant species—the Kurumba, for example, discovered that the leaves of even Eupatorium odaratum, the pernicious recent introduction into the forests of the south, could be used as a poultice for some varieties of wounds. The density of the cultivated plants being small in comparison with the natural vegetation, grazing by wild animals was entailed no significant loss.

The most widespread method, in fact synonymous with the system of shifting cultivation, is the Slash and Burn. The knowledge base has to include an appreciation of the behaviour of fire in given conditions of wind and methods of controlling the burn. The "cooling" of the soil thereafter was a matter of fine judgement. The choice of planting material was delicate and important. It was also necessary to have a prescient knowledge of the time when a site should be given up and another chosen. An instinctive appreciation of wild life behaviour including their migratory patterns and preferred browse and graze species was necessary in order to minimize the incidence of crop raiding by them.

It is not easy to realize, but none the less a fact that, for mankind, the invention of Slash and Burn cultivation was a technological innovation not inferior in significance to any subsequent "Revolution", 'green' or otherwise. Given the man-short, land-abundant conditions in which it was invented, it was a means of maximizing the productivity of human labour. The nutrients stored in the tissues of the plants burnt down were available for the cultivated varieties in addition to the soil nutrients. There was a rotation of cropping because of the change of site after a period of time, nature repairing the damage caused by the invasive technique. Tribal wisdom discovered the optimum time cycles required.

Mankind had to pay a price for the advantage gained; the conditions of slash and burn were such as to favour the multiplication of the Anopheles gambiae mosquitoes and the associated spread of malaria. Livingstone 1969 has provided strong evidence that the maintenance of the HbC (sickle-cell) haemoglobin represents a balance of polymorphism through which human populations adapted to the malarial parasites in response to the falciparum malaria caused by Palsmodium falciparum. Some African populations show a different haemoglobin pattern in response to another kind of malaria caused by P. Malariae (Beals, *et al.*, 1971; Kottack 1974).

Evidence in India indicates that tribal communities developed a complex culture based on this method of cultivation. Great care was taken to ensure that no permanent damage was done to the edaphic conditions available in any site. On the other hand, tribes like the Kui Khonds of Orissa took care to stack felled materials along the contours to reduce soil erosion due to tillage (Nayak *et al.*, 1990: 114). Fruit trees were invariably preserved during the slash and burn operation. Shifting cultivation vested no permanent ownership of land on an individual or family. Most importantly, for the communities practicing it, shifting cultivation was never the sole means of livelihood; the invasive technique was thus kept within limits well understood in the cultural traditions of the community. (For a detailed analysis of tribal practices of shifting cultivation, please see the article under this heading in Menon, T.M. (ed.) 1993: Awaiting Publication).

Thus, shifting cultivation can be a stable, productive and "sustainable" method of co-existence of man and animal in a forest habitat if the following are satisfied:

(a) The land-man ratio is favourable in the sense that man has a very wide choice of areas which he can cultivate, leaving the rest alone;
(b) The shifting circuit leaves a sufficiently long interval before a site is cultivated for a subsequent turn;

(c) "Pollutants" in the form of chemicals, exotic plant materials, etc., are not brought in—a condition which is automatically satisfied when the communities practicing shifting cultivation were isolated and left free from the zeal of "extension" agencies;

(d) The proportion of the land remaining uncultivated to the area being cultivated was so great that wild life had sufficient forage without recourse to crop raiding; and

(e) The cultivation entailed no permanent structures like dams and roads on the area shifted from during any part of the cycle.

In a recent study (Muraleedharan & Sankar 1991), the Kerala Forest Research Institute, Peechi evaluated the agro-ecosystem of some tribal communities against five properties of such systems, *viz.*, productivity, stability, sustainability, equitability and autonomy. The study found that although the productivity of individual crops is low, the total productivity of the crop mixture under shifting cultivation corresponded with the average yields elsewhere in the State. It was also reasonably stable during the usual three years before the site would be shifted from, the decline being the low availability of biomass for burning during the second and the third years.

An analysis of the vegetation in plots of varying fallow length revealed that when abandoned after a cycle, coppice growth was plentiful, the build-up of vegetation increasing with the length of the fallow period. The redeveloping forest resulted only in secondary forest with pyrogenic tree species like *Dalbergia latifolia, Grewia tilafolia, Emblica officinalis*, etc. The absence of significant differences in herbaceous and grass growth among plots of varying fallow length indicated that the area was not yet prone to be converted irreversibly into an unproductive grassland.

Equity was scrupulously maintained both because members of a hamlet are relatives, surpluses being shared, as well as by changing the allocation of fields to each family so that each is afforded an equality of access to the more desirable sites. It was "autonomous" in the sense that the community could carry out the cultivation with their own resources and family labour, and in that, most of the produce being for local consumption, the production patterns were not determined by external market forces or preferences.

The most significant finding was that compared to settled cultivation adopted by the adjoining tribal communities, shifting cultivation was more "efficient" in the ratio of the output in terms of energy to the input measured in the same terms. It was 19.50:1 in the case of shifting cultivation, whereas in settled cultivation without the use of animal power or chemical fertilizer, it was 16.26:1; and in the case of settled cultivation with the use of both these inputs, it was only 12.85:1 (Muraleedharan & Sankar, 1991).

Under such conditions, the extent of damage done by man to the eco-system in the course of shifting cultivation was minimal and reversible; centuries of such practice by tribal communities the world over did not destroy the forests. Only the settling of cultivation did so inexorably and irreversibly! It is also clear that tribal communities themselves were aware of the fragility and sensitivity of the system—hence the various observances, taboos and practices intimately interwoven with their cultural attributes, and forming part and parcel of their knowledge and belief systems. Hillel (1992) pointed out that, though according to the Food and Agricultural Organization, shifting cultivation is responsible for 35 to 70 per cent of deforestation, this does not permanently destroy the forest cover, provided that the affected plots are given adequate time to regenerate. "In an eco-friendly slash-and-burn system, the burning would go on in the kitchen, with ash are turned to the plot. People cause most trouble because of their greed, stupidity and unwillingness to use existing knowledge" (Pirie, 1992).

Transition from Shifting Cultivation

A combination of circumstances such as the vesting of ownership and central control over forests in the Government, mono-culture plantations, non-tribal encroachments, the intrusion of market forces and economic exploitation, and the increasing strictures of departmental control over the use of forest resources, including the greater emphasis on the protection and preservation of wild life, had the effect of rendering shifting cultivation a "nonviable" alternative to forest dwelling tribal communities. Iyer (1937) reported that "permanent cultivation" and "wet cultivation came into existence as a result of restrictions imposed by the Forest Departments in Shencotta and Quilon Divisions". Communities like the Malayarayar, Muthvar, Kanikkar, etc. have already developed settled cultivation partly in imitation of and mostly in defence against the incursion of "settlers". Fully forest-resident and forest-dependent communities like the Kader and the Kurumba are increasingly finding agriculture a "non-viable" alternative. They have become "Proletarianized" as wage earners under various government departments and their contractors. They also continue to find occupation as collectors of minor forest produce.

It was not long before such interferences resulted in edaphic consequences such as erosion and soil loss and the drying up of water courses. The forest soils were exposed; in the southern States, e.g., Kerala, the process of laterite formation ensued. The intensification of tillage accelerated these changes; entire hill sides became denuded of foliage, scarred with gullies, with laterite patches not supporting even grasses. What little vegetation was left behind became the prey of annual fires sweeping the hill sides. Blame was attached to the tribal practice of shifting cultivation. Brain-washed by concepts of agriculture seeking to maximize gross product per unit area of land, experts condemned shifting cultivation as inefficient and cause of soil erosion, forgetting that it was precisely the intensification of tillage rather than the shifting character of the cultivation that was responsible. Foresters as well as Agronomists looked on shifting cultivation as the "unforgivable sin".

Settled Agriculture

Historically "Revolutions" in agriculture, from the Agrarian Revolution in Europe a couple of centuries back to the "Green revolution" the world over a couple of decades back, were characterized by the objective of maximization of output of chosen produce over unit area of land devoted to its cultivation. The main thrust of agricultural research was the evolution of planting materials with the maximum yield potential in a given situation; if possible, the "situation" was altered to make it suitable for the expression of the genetic potential of the planting material so evolved, e.g., by providing irrigation, intensive tillage, etc. If the planting material had little resistance to pests and diseases endemic or epidemic in the situation, then the use of chemicals and pesticides was recommended. Other plant species were treated as weeds, to be kept down, preferably eradicated, by tillage and other means. Tillage was an "invasion"; the more intensive the technology used, the more severe the intensity of the invasion.

The intensification of this process led to the irreversible loss of forests and wild life. The application of engineering and other technologies to change the nature of the land and the edaphic characteristics of the environment led to a "vicious" circle; the more the alterations, the more continued the requirement to apply technology to maintain the changes. Together with this, the human biotic interference led to the disappearance of several life forms and the emergence and increase of several others, again necessitating continued application of the interfering processes. These consequences are now so well known that they form the standard contents of most writings on the subject. Agricultural experts are now realising the wisdom of "going back to Nature" in the sense that the minimum disturbance of natural balances is the key to sustainability of habitat and productivity in a given environment.

Beleated Concern for Wild Life

Even though it has been long known and acknowledged that wild life is one of the most invaluable assets of a country, it was only in the 1980s that the concern became "Natural". Stringent legal provisions were enforced to ensure the protection and preservation of wild life. The "Project tiger" was an outstanding example of an effort. This has however introduced a third element in the equations determining appropriate land use.

Some Kerala Cases

The tribal communities of Kerala represent a spectrum from those in the hunting gathering stage, through shifting cultivators to advanced plantation economies. Even from the early thirties, the Forest Departments in the territories comprised in the Kerala of today had been conscious of the Government's vested interests in the Forests and the need to maximize revenue from them. This conflicted with the tribal interests of utilizing a vast area of land in shifting cultivation circuits, and to the Departmental outcry against it in the interest of "conservation" of the tree growth of commercially valuable species like teak.

The Cholanaikkan have been described as the "cave dwellers" of Kerala. They are found in the dense ever green forests along the escarpments of the Kundah Range of the Western Ghats where the natural caves afforded the most convenient abode. These are approachable only through steep climbs, thus protected from wild animals, especially elephants. The forest is so bountiful that no competition for natural produce among the species including the humans arises. The humans have learnt that they have nothing to fear from elephants if they are left alone and if the human "freezes when rarely and actually confronted with the "Pachyderms". Cholanaikkan sometimes trap small animals and catch fish, but run the risk of being killed and eaten by bears and the other predators. But so satisfactory is the technological equation they have evolved to meet their requirements from the habitat that they find time to make toys from beeswax and to worship the deities, among whom they include the elephants and the tigers.

The Kurumba of Attapady and the Muduga with whom they inter-marry have a myth of creation according to which the elephant was the first creature to emerge at the hamlet Anavayi (meaning "the Elephant's mouth" or "the way of the Elephant"). Kurumba folk followed. They accord to the elephant respect and veneration in the form of an avoidance relationship as for an elder. The hamlets are generally sited on the brow of a hill, and are built or wattle plastered with mud, merging with the natural surroundings. Until recently, they adopted a peculiar system of shifting cultivation. In their wet evergreen habitat, it was not easy to maintain a fire and adopt the "slash and burn" method; instead they used to "scratch" out some of the natural vegetation by digging with the "Koonthali" (pointed stick). They would then simultaneously sow the seeds of red gram, coarse grains and maize. These would be harvested in due course in the turn in which matures; the produce, along with the wild roots, tubers and fruits constituted species, including the pernicious recently introduced weed, *Eupatorium odaratum*, as a poultice.

Under such a system of farming, no cultivated crop showed prominently enough to attract wild life, especially the elephants. The density of the cultivated vegetation was so low that the incidence of an occasional damage by a stray animal was minimal. The most significant of their usages was one narrated to me by an old headman of the Muduga in 1959. They would keep collections of produce including waste like plantain pseudo-stems along the paths frequented by elephants, but leading away from the hamlet and plots currently under cultivation. The headman called this the share due to their elder brother and the lord of the forest, viz., the elephant. If this was promptly and liberally offered, then there was nothing to fear of crop raiding by the elephants. This practice must have been the result of the close observation of the behaviour of elephants, a study of their migratory paths, their preferred browse and graze species and the manipulation of the herd away from the cultivated sites. As far as I know, there have been no reports of elephant attacks on their traditional hamlets.

The Kattunaikkan (Kadu Kurumba in Tamil Nadu and Karnataka) are an ancient tribe found along the Kerala-Karnataka-Tamil Nadu trijunction in Wyanad, in the deciduous forest with high density of wild life, especially elephants. These forests were the sites for the famous Khedda operations for the capture of wild elephants; the Kattunaikkan were the unsung "heroes" of these operations, driving the herds towards the palisades constructed for the purpose, and trapping them once they were safely inside. The Kattunaikkan then served as the expert *mahouts* who tamed and trained the captives. They are still preferred by the forest department for employment as *mahouts*. Traditionally artisans for the neighbouring tribal communities, their main source of livelihood is the collection of minor forest produce, with occasional shifting cultivation.

The Malassar on the eastern aspect of the Parambikulam Range are also preferred for employment as *mahouts* because of their knowledge of the ways of the elephants. They occupy one of the richest wild life habitats, amidst moist deciduous and evergreen spread and other areas of the Parambikulam Áliyar Multi-purpose Project have submerged several square kilometers. The present Malassar settlements are resilted ones, along the banks of the reservoirs. Malassar have sporadic cultivation within the limits of the hamlets, in the immediate vicinity of their huts. No case of crop raiding by elephants has been reported but wild pigs are a pest. Perhaps the plentiful availability of forage in forests and the rigorous protection enforced in the sanctuary areas has reduced the need, as far as elephants are concerned, to raid crops raised by the humans.

The Kadar, living in the interior of the evergreen side of the Parambikulam area, have had quite different experiences from the elephants. They are an ancient tribe classed by some authorities as Negrito in origin, and hence among the earliest inhabitants of India. This habitat was the source of the elephants for the erstwhile Cochin State; the elephants used to be captured by the "pit method". Kader used to be employed as the "pit watchers" who reported the falls, and were employed in the hazardous "out-pitting" operations, when the captive would be taken out with the help of trained elephants and driven to the *kraal* for the training process. Even before Independence, the then Cochin Government tried to settle them and "wean them away" from their traditional hunting, gathering and shifting cultivation modes. Elephants persistently raided the crops because they were right in their migratory paths. The tribe continues to depend on the collection of minor forest produce and employment under various departmental agencies.

About 1979-80, elephants attacked a small settlement at Orukombankutty near Chalakkudy river where new houses had been constructed. The terrain is comparatively open as the tree growth had been extracted long ago and replaced with reeds and grasses, thus a rich feeding ground for elephants. The elephants demolished all the new houses and killed some humans including a Forest Guard. The settlement had to be abandoned by the humans. I visited the site about a year after the incident and found the elephants fully in possession and disdainful of occasional human intrusion; the herds were calmly browsing showing no intention to charge.

The Kanikkar trace the origins of their clans to the finding of the carcass of an elephant. Iyer added that in Kulathupuzha, "if a man clears some lands where elephants roam about, he decides to make an offering to the deity Pulichavu after harvest. If he had no trouble, men hull twelve padis of paddy and the rice obtained therefrom is cooked and placed over a leaf. The medicine man then prays. . . . "Oh! Anavila Pulichavu,. . . I make the offering as I promised . . . accept with gladness and drive away all harm!". In 1937, Iyer had reported that the "Kanikkar are still nomadic agriculturists. In the uplands cultivation is migratory. In the lowlands, cultivation is permanent. On the hills, rice is grown in the dry system . . . individual property is not recognised. The Kanikkar to the north of the Vamanapuram river have wet cultivation. . . . Wet cultivation came into existence as a result of the restrictions imposed by the Forest Department on "Taugya" cultivation in Shencotta and Quilon Divisions" (Italics supplied). Now, however, except for isolated communities in the interior forests, Kanikkar have become settled cultivators. The forest has disappeared from the areas so settled.

Some General Observations

Once a human community in sufficient numbers takes to settled cultivation, the loss of forests and the elimination of wild life becomes inevitable and irreversible. Kanikkar and Malayarayan, compelled to give up shifting cultivation because of Government policies, adopted modes and crops which can no longer subsist in a forest environment. The Kattunaikkan case shows that the increase in crop intensity and vegetation density itself attracts crop raiding by elephants. The human reaction is to massacre, often by the most brutal methods, the animals involved.

Alternatives could take the form of engineering works like trenches and ditches, fences, simple or electrified, etc. These are excessively costly to establish and maintain. Humans trend to cross them in their pursuit of more areas to cultivate, thus inviting crop raiding; then they raise louder complaints than ever of crop raiding by elephants! Resiting of human populations away from forest habitats rich in animal life entails cultural changes apart from the costs involved. The capture and evacuation of elephants to sanctuary areas is also costly. Humans tend to get back to their original locations; so, it has been reported, do elephants.

The Three Considerations

The above discussion would indicate that there are three elements the equation of which vitally determines the sustainability of land use in a forest ecosystem. They are the humans and their needs, satisfied by a "technology" they have developed whereby the habitat, a "given", is altered and manipulated; the animals, who threaten and are threatened by too intensive and invasive technologies developed by the humans, and the forest itself, a unique balance achieved by Nature with the global conditions. The objectives of continued occupation by humans of forest land, replacement of their shifting cultivation by settled agriculture and the preservation and protection of forest and wild life are NOT simultaneously possible; at least one of them has to be sacrificed. Alternatively, systems of agricultural land use have to be developed which retain some of the values of shifting cultivation, sacrificing at least to some extent the yield maximizing strategies adopted in settled agriculture.

Man as an animal, though no longer generally a denizen of the forests, still depends on them; many humans continue to be forest-dwellers. The maintenance of the forest is not merely the planting and protecting of trees, but a total preservation of the inter-relationship of many life forms, in which the humans are an important but not the only component. However, avidly a nation pursues a policy of rapid industrial growth, whatever socio-political power structure and interest systems it generates, the fundamental equation of forest, man and animal has to be on the basis, NOT for the maximization of gains of the dominant section of Man in his *avatar* as Captain of Industry, but the minimization of losses of forests and animals and the protection of the interests of the forest dwelling communities, often among the poorest of the poor who are the repositories of value systems significant for mankind.

The technologies adopted by these communities have stood the test of time over centuries before the intensification of exploitation of land and forest became so great that more invasive methods were developed. These technologies were part and parcel of the life styles, beliefs and value systems, rather than systematized formulae and manuals of procedure. They subserve the objective of a yield that can be "sustained" without unduly destroying other life-forms or altering the basic characteristics of the soil by excessive interference such as irrigation or chemical treatments. Conservation of native gene pools as potent sources of resistance to pests and diseases is given high priority. Intensity of tillage is reduced to avoid soil and moisture loss.

Preservation of Life Forms

The protection of wild animals has always been professed as a policy priority. However, as the density of human populations increases and the margins of cultivation extend, conflicts arise. They are especially marked in the erstwhile forest areas which humans occupy. Under the "Dig and Scratch" system as adopted by the Attapady tribes, crop raiding by animals hardly ever happened; the density of the cultivated plants was not significantly higher than that of the native vegetation. The problem was not very severe even under the "Slash and Burn" method, given the original extensive circuits; the area under cultivation was minute when compared with the area from which the animals could naturally forage and the cultivation circuit did not generate a preference among the animals for cultivated vegetation over the natural growth.

Crop raiding becomes a serious problem for communities when the cultivation circuits are curtailed. Under the pressure of increasing population, the community needs more food production; they try to raise it by intensifying the tillage and increasing the crop density. Simultaneously, animals are also under food pressure; their ranges have been reduced by human interference in the form of mono-culture plantations, nonforest land use, etc. The tribal people are prevented from culling out the crop raiders because of the rigours of the Wild Life Protection laws.

As settled cultivation extends to the margins of the forests, animals develop preferences for the cultivated plants. Unlike tribal people who "know" animals and avoid "personal" confrontations with them, the "settlers" try to drive them away. A few humans are maimed or killed by the enraged animals; a hue and cry is then raised through all the mass media. Fences are put up, some electrified, and trenches dug at huge cost and the animals driven back. The humans then extend their encroachment beyond these defences, often breaching them; the cycle is repeated. The technocratic alternative is to create more and more

"Reserves" within the forest. Each such reserve pushes back the tribal people a little more into distress and imposes more restrictions on them; poaching and encroachment by more resourceful communities continues.

There are also several life forms including micro-organisms which are vital for the preservation of forest ecological systems. The equation of forest, man and animal has to include them also, if it is to result in beneficial interfaces. The impact of industrial technology, such as the dumping of wastes in the forest or the drainage of effluents, as from a "High-Tech" spare Science facility in the forests around the Mahendragiri Peak of the Southern Western Ghats, has to be minimized.

Edaphic Appropriateness and Biotic Equity

The essence of the sustainability of a form of land use in a given area has to be: the use, including the choice of crop if the use is agricultural, which is "edaphically appropriate"; at the same time, the life forms which are endemic must not be too severely altered or eliminated in order to favour the introduced forms.

In Nature, edaphic appropriateness is enforced by survivability of the life forms which stabilize in a region. Human technology alters this condition by changing the edaphic characteristics, e.g., by providing irrigation in an arid area, ameliorating soil conditions by physical and chemical manipulations such as tillage and fertilizer applications, etc. These alterations require continued application; otherwise the natural conditions tend to react in various ways to revert to as close an approximation to the original conditions as possible, often at a lower level of vegetative wealth than before; the ultimate low level could be that of "desertification", not merely a theoretical possibility!

Human interference may initially be triggered off by the choice of a "crop". Eucalyptus may replace a natural evergreen forest (e.g., the Lady Smith Reserve in Wayanad) or teak or tropical pines may be raised cutting down a sal (*Shorea robusta*) forest. Paddy may be planted in steeply sloping terrain, the soil requiring constant protection by bunds, and watered through irrigation works, requiring more interference with Nature and alteration of edaphic characteristics. Under such conditions, natural vegetation will become "weeds" to be eliminated or kept under control in various ways. All these are the biotic consequences of "edaphic inappropriateness".

Biotic Equity is a more complex concept. No conservationist, however committed, will probably argue for the conservation of the plague bacillus or the small pox virus. At the other extreme, no Finance Minister, however concerned with the maximization of profits for industrial entrepreneurs, will seriously urge the extermination of the tiger or the elephant, even though most of his priorities may lead to such an eventuality. "In practice", the problem is to identify those life forms and their inter-relationships and to ensure their best balance such that their interconnectedness with human communities ensure the greatest survivability of all.

"Good" Shifting Cultivation

We have been so brain-washed by the outcry against shifting cultivation as to automatically identify it as "BAD" cultivation. I believe that we have to redefine "BAD" cultivation as "agricultural practices edaphically unsuited to a particular area during a particular period of time, and/ or—(usually 'and' rather than 'or')—biotically inequitable in a given complex". Viewed thus, it will be "settled agriculture" that will be "BAD" in several locations, if the conservation of forests and the preservation of wild life species be among our value objectives.

Shifting cultivation, if the shifting circuit is sufficient well spaced out in area, and, more importantly, time, could be the "sustainable" alternative in several locations. The Kurumba practice of "dig and scratch" seems to be an ideal solution; it harmonises the needs of the humans in the midst of a wild life population which has to be absolutely protected. Because the tillage involved is minimal, the soil is not unduly disturbed as to cause erosion even in the steepest slopes. Natural vegetation is NOT treated as weeds and is disturbed only to the least extent. The practice is edaphically appropriate and biotically equitable.

The Problems of Productivity

This is another concept the relativity of which is often not realised. "Productivity" is after all a ratio; both the numerator and the denominator depend on the definitions adopted. For example, the objective function to be maximized could be gross tonnage yields or other preferences such as calorific value, taste size, etc. It is only when the former object of maximizing gross tonnage yields is assumed that edaphic appropriateness and biotic equity are sacrificed. If we consider these values as desired outputs of the use of resources, "productivity" takes on different meanings.

A globally generalized model of the inter-relationship of resources, technology and output could take the following form:

Definition

1.1. "Technique" is an operation whereby, by itself or in combination with other operations, a desired output is obtained in a given condition determined spatially and temporally by circumstances.

1.2. "Technology" is a set of techniques

1.3. "Appropriate Technology": Technology is appropriate when it accomplishes the production of the desired output in the given condition with the minimum expenditure of those items of input which are the scarest or the most desired to be preserved or conserved.

1.4. "Adequate Technology: Technology is adequate when, in the given conditions, the output is sufficient, both quantitatively and qualitatively, to meet the requirements of the community concerned for it.

1.5. "Stability Conditions": denote the set of conditions required for the continuation of a technology which is appropriate and adequate.

These definitions lead to a proposition which can be stated thus:

$$(C) * (T) \rightarrow (Q) \qquad ...(1)$$

where

"(C)" is the set of circumstances "here and now", in which a human community lives;

"*" is an application or use of:

"(T)" is the Technology set;

"(Q)" is the set of the desired output.

As all human communities are omnivorous and gregarious, and apply technology to obtain a desired output from a given endowment of resources, the above proposition can be universally quantified thus:

$$(x)\, Hx,\ \{(c) * (T) \rightarrow (Q)\} \qquad ...(2)$$

In other words, all human communities apply technology in their circumstances to obtain a desired output.

It is possible to recursively define the operation "*" as:

$$* \text{ def} === f\, \{(C), (T), (Q)\} \qquad ...(3)$$

viz., the operation * is a function of the sets C, T, and Q.

Definition 1.3 imposes an "efficiency condition" on (*C*). Inputs are comprised in the given (*C*) and constitute the resources with which the community is endowed. Rational behaviour would require the deployment of these resources in such a way that the minimum expenditure of those items which are in the shortest supply or to which the greatest value is attached is incurred for the production of the desired (*Q*). Efficiency is thus a criterion for the "appropriateness" of technology and is "locally determined" in the sense that the productivity of a specific item of resources in (*C*) is maximized.

An analysis of the "slash and burn" method of shifting cultivation will make this clear. Where land is abundant and population (labour availability) is low, efficiency required the maximization of the productivity each unit of land employed for the production "food". So long as the initial conditions, viz., extensive land availability and low labour availability subsisted, the long periodicity of the shifting cycle ensured the long term restoration of the productive balances. The illustration also serves to bring out the conditions in which an appropriate technology can cease to be such because of change in (*C*). Curtailment of the shifting circuit, e.g., by Government interference or competing use of land, will reduce the availability of land. Simultaneously, reduction of land availability will render the increased labour availability an unemployed surplus. The marginal productivity of both land and labour will reach a critical point where further increase in production will require change in the technology involved.

Corresponding to the above, definition 1.4 imposes a "sufficiency conditions" on (*Q*) in relation to (*C*). Even when a technology is appropriate in the sense that it does produce (*Q*) minimizing expenditure of certain elements in (*C*), (*Q*) may not be sufficient either qualitatively or quantitatively to meet the requirements in (*C*). In other words, given a constancy of (*C*), (*T*) and (*Q*) may not be enough to satisfy the demands, physical and cultural, of the community. This condition is value-loaded in that cultural preferences and desirability criteria are determined by extra-economic considerations: "Man does not live by bread alone"!

Stability Condition

Definition 1.5 postulates that when a (*T*) is both appropriate and adequate, there will be a set of conditions that will make it continue. Conversely, when the conditions are such that a (*T*) is rendered either not appropriate or adequate, then stability will not ensue. Communities react to destabilizing factors by innovations in technology. All innovations are qualitative changes and have cultural consequences ranging over (*C*) and (*Q*).

Lessons from Tribal Experience

The experiences of tribal communities in India indicate the basic level at which the interactions of resources, technology and output result in a stable equilibrium which is edaphically appropriate and biotically equitable. The inadequacy of the technologies adopted by them revealed when the resource endowment is altered by Government polices, external influences, etc. and/or when the needs of the community change because of quantitative increases in population or qualitative change in preferences and life styles.

When globally generalized, the human condition can be seen to be the analogue of the tribal condition, raised many times by the power of human numbers, diversity of needs and complexities of technology. However, the basic relations among resource endowments, technologies and desired output remains true, and the challenge continues to be the conservation of those elements in the first category which are scarce, irreplaceable or excessively fragile.

Rabindranath Tagore, with the prophetic vision of the humanist poet wrote:

"Mother Earth has enough for the healthy appetite of her children and something extra for rare cases of abnormality. But she has not nearly sufficient for the sudden growth of a whole world of spoiled and pampered children. Man has been digging holes into the very foundations, not only of a whole world of spoiled and pampered children. Man has been digging holes into the very foundations, not only of his livelihood but of his life" (in Elmhirst, 1975).

REFERENCES

Agarwal, Anil (Ed), 1992: *The Price of Forests*, Centre for Science and Environment, N. Delhi.

Beal, R.L., Heijer, W. and Beals, A.R., 1971: *Introduction to Anthropology*, Mac Millian, N. York.

Hillel, Daniel, 1992: *Out of the Earth: Civilization and the Life of the Soil*, Free Press, London.

Iyer, L.A.K., 1937: *The Travancore Tribes and Castes*, Vol. I., Government Press, Trivandrum.

Kottack, C.P., 1974: *Anthropology: the Exploration of Human Diversity*, Random House, N. York.

Livingstone, F.B., 1969: "Gene Frequency Clines of the B. Hemoglobin locus in various Human Populations and their simulation by Mendel involving Differential Selection", *Human Biology*, 41:223-236.

Menon, M. (Ed), 1993: *The Encyclopaedia of Dravidian Tribes*, (awaiting publication), International School of Dravidian Linguistics, Trivandrum.

Muraleedharan, P.K. and Shankar, S., 1991: *Human Ecology and Socio-Economic Interactions in Tribal Communities of Attapady*, Kerala Forest Research Institute, Peechi.

Nayak, R. *et al.*, 1990: *The Kondha—A Handbook for Development*, Indian Social Institute, New Delhi.

Perie, N.W., 1992: "Ground rules", *Nature,* Vol. 357, 14, 372-374.

Sukhumoy, Chakravarthi, 1992: "Sustainability: the Concept and its Economic Implications in the context of India", in Agarwal, Anil (Ed), 1992, *Op. cit.*

Tagore Rabindranath, 1975: In Elmhirst, L.K., *Poet and Plowman*, Viswabharati, Calcutta.

12

MICRO-PLANNING APPROACH AS AN EFFORT TOWARDS SUSTAINABLE DEVELOPMENT
Experience from Tribe Inhabited South-West Bengal

R.N. CHATTOPADHYAY, D. BHATTACHARYA,
CHANDREYEE DAS and TAPATI DASGUPTA

Introduction

MICRO-PLANNING is synonymous to local level planning which arises as a result of local, often isolated, initiatives without special institutional set-up or links with the national level. Hence, the objective of such a planning mechanism means a design of more realistic projects and programmes taking into account the local potentials and constraints. Its functional efficiency, however, could be strengthened by boosting the mobilization of local resources and increased people's participation in development.

The centralized or the top-down approach towards planning often had faulty implications in the sense that the design was not need based and the impact of such an exercise did not trickle down to appropriate areas. This, in effect, led to a lopsided and contemporaneous development. Microplanning could thus be considered as a corrective and efficient process for development at microlevel. A qualitative and quantitative assessment of the local resources like land, human resource in terms of skill, adaptability, social set-up, economic standing etc. coupled with an analysis of the dimensions of the area specific problems and constraints and the aspirations of the local people help in the formulation of need based plans. The strategy for microplanning would involve the ways and means to utilize the local resource potentials, especially the latent resources of unused and under used labour, skills and knowledge and to increase motivation in the individuals to produce better. This would not only curb the local problems but would generate a sustainable, self sufficient path for development.

The four decades of planning in our country have definitely opened up different avenues of development but the backward regions in many cases have maintained their *status quo* or have even further deteriorated in terms of socio-economic well-being. Though several programmes based on the area approach or ethnicity approach for backward area development have been implemented, in reality no significant benefit has accrued to this section of the society. Hence, the requirement is to design local level plans and programmes within the broad framework of the national policies for growth and development.

The attributes and benefits of microplanning as conceived by planners could be as the following:

(a) bidirectional flow of information during planning.
(b) people's participation in resource management and problem solving.
(c) site specific and need based plan preparation
(d) fixing realistic objectives after considering aspirations and expectations of the local people,
(e) participatory linkage between user and planner,
(f) better identification of local resources,
(g) better chance of success, and
(h) ensure grass-root monitoring and evaluation.

Keeping in mind, the set of features discussed above the present study aims at planning sustainable development for a backward tribal region through the micro planning approach.

The present study describes the steps in the formulation of a microlevel plan for a typically backward tribal region of Jhargram in the district of Midnapore, West Bengal. The action plans and programmes drawn would give an indication of the effectiveness of a microplanning exercise in evolving a self generating process of development.

The tribals in India are about 8 per cent of the total population. Percentage-wise the figure seems to be very small, numerically it reaches the height of 54 million. This chunk of the autochthonous populace has always been passing through the economic hardship of life and facing deprivation in the availability of economic, social and infrastructural facilities. They neither could be a part of the mainstream society nor could they utilize the fruits of development to an expected extent.

After independence the National Government tried to deal with the ardent problems of the oppressed tribal people for their overall development, a host of programmes have been initiated. The tribals were provided with Constitutional Safeguards in order to prevent them from further deprivation. Tribal development plans and subplans had been announced and a good deal of money had been spent for achieving the required goals but in reality no substantial development could be traced.

The present study area consisting of seven villages of the Jhargram block in the district of Midnapore, West Bengal is the habitat of such a down trodden section of tribals. The different aspects of their bleak existence has been portrayed through information in the form of factual data in the following section.

Case Study: A Brief Review of Existing Features of the Field of Study and Analysis of Problems

The area under our research embraces a stretch of land amounting to 630 hectares. Nearly 60 per cent of the land is under cultivable use, the yield is very low due to unsuitable soil and land characteristics. The cultivable land is mostly rain-fed. The forest coverage under the jurisdiction of the villages is a little less than 5 per cent of the area. A chunk of land is left waste due to lack of concern and investment. The landuse distribution has been provided in Table. 12.1.

TABLE 12.1 : Landuse characteristics

Landuse Type	*Area in Hectares*	*Percentage*
Total land	629.95	100.00
Forest	28.75	4.56
Irrigated land	0.0	0.0
Unirrigated land	368.55	58.50
Cultivable waste	14.95	2.37
Built up, Waste land etc.	191.43	30.39

The total population of the present area is 980 with a sex ratio of 980 females per 1000 males and the decadal population growth is almost 23 per cent. Population density figure is 156 persons per sq km. Tribals count for more than 97 per cent of the total population (see

Table 12.2). A number of tribes inhabit in the seven villages under our study amongst which Santals comprise the most populous ethnic community. Over 80 per cent of the total tribal population belong to this group. The following dominant tribe is Lodha. Apart from these two communities there are some other Scheduled Tribes like Munda, Baiga, Mahali, Savar and Bhumij. They constitute a considerable size of the population in the neighbouring regions. The composition of population is given in Table 12.3.

TABLE 12.2 : Demography

Category	*Persons*	*Percentage*
Total population	980	100.00
Male	495	50.51
Female	485	49.49
Total population (Tribal)	851	100.00
Male (Tribal)	423	49.70
Female (Tribal)	428	50.30

TABLE 12.3 : Village level composition of population

Village Name	*Population*	*Santal (ST)*	*Lodha (ST)*	*SC*	*Other Castes*
Kalaboni	169	75	88	—	6
Basantapur	47	21	7	12	7
Chinchuurgeria	86	86	—	—	—
Pasro	338	248	—	70	20
Piyalgeria	134	134	—	—	—
Mahulboni	70	56	14	—	—
Chhoto Lodhasuli	136	122	—	14	—
Study Area	980	742	109	96	33

Most of the tribals especially Santals are agriculturists though their livelihood largely depends on the forest-based economy (Table 12.4). Lodhas, Savars etc. earn a living mainly through gathering minor forest produces. These tribal communities are different from the former in the socio-cultural aspects. The socio-economic condition of the tribals irrespective of their ethnic origin is found to be very low, but that of the Lodhas need a special mention. It is worst. Their poverty-compelled involvement in criminal offence have created a social stigma against intermixing with them.

The economic activities carried out in our tidy area belong directly to the primary sector or its allied sectors. The major occupation is either agriculture or forest related activities amounting to negligible income. Agriculture of course predominates over other activities and in all villages it is taken up as the major occupation. Paddy is the major crop.

TABLE 12.4 : Scheduled tribe workers in different activities

Total	*Forest based activities*	*Agri-culture*	*Agricultural labour & other wage earners*	*Rural industries and crafts*	*Other services and activities*
340	62	148	90	16	24

The number of landless and leased landowners constitute a chunk of the persons involved in the agricultural sector. This definitely indicates the strained economic well-being of a large section of the inhabitants. Marginal farmers override all other categories of cultivators and is closely followed by small farmers.

Animal husbandry or livestock rearing is another source of livelihood allied to agriculture though earnings from this source are too meager. Most of the families in these villages rear cows, buffaloes, goats, sheep and pigs. Poultry birds are also nurtured but not in an organised manner. The average earning from livestock per household amounts to Rs. 100 per month.

The forest related activities include, collection of some selective non-timber forest produces (NTFP) and their processing. Illegal felling of trees for timber had been a prevalent practice but government orders and the programmes of Joint Forest Management (JFM) have acted positively in stalling this activity in this region.

In the secondary and tertiary sectors of the economy the inhabitants of these villages contribute very little. The rural and cottage industries sector is still in its primitive stage of operation. They mainly produce minor items like leaf plates, baskets, broomsticks, mattresses, etc. Distillation of country liquor is one of the prevalent cottage industries. The characteristic feature of a tribal community is that the women participation rate is the same as that of the male participation rate. In a typical tribal household consisting of four to five members the lead pair constituted of a male and female member is involved in some economic activity. Thus considering the fact that out of five persons two are engaged as workers, the overall participation rate turns out to be around 40 per cent. In our study area the percentage of workers is estimated to be 39.99 ~ 40 per cent.

The study area portrays all signs of low economic development. As revealed from statistics it shows distinct signs of impoverishment. Poverty stricken environment, malnutrition and poor civic conditions characterize the villages as well as its inhabitants. This is mainly due to the meager earnings of the people which again results from the former. About 80 to 85 per cent of the tribals reside below the poverty line earning less than Rs. 500 per month per household. Considering an average household size of 5 to 6 persons the hopeless condition of this class can very well be imagined. The income pattern in the major economic activities is shown in Table 12.5.

TABLE 12.5 : Average earnings from major economic activities

(*Rs./month*)

Agriculture				*Forest*	*Live-stock*	*Cottage industry*	*Other services business*
Marginal farming	*Small farming*	*Large farming*	*Laboures*				
300	550	1500	150	250	100	250	1000

The tribal economy is basically production oriented. A major portion of their produce is used for self consumption while the remaining portion is transacted either in weekly '*hats*' or small daily markets, which are often located at sizeable distances. The large daily market, which is the Jhargram urban market, hardly leaves any scope for their regular penetration. This necessitates the diversification of their products to cater to the needs of such markets and identification of proper channels of marketing.

The expenditure pattern of the tribal households shows that a major portion of the income of the tribal people are used for consumption i.e., mainly food. Moreover, alcohol addiction and lack of propensity to save, lead them to extra expenditure. This further creates a situation where, to meet the basic needs of the family, a person has no other way but to go for borrowing money from landowners, businessmen or rich people of the vicinity. Usurious activities of these money lenders make their economic condition even worse.

Most of the villages under our survey suffer from multiple problems. Illiteracy, malnutrition, unhygienic state of living coupled with poverty portray the inhuman condition of the people. The villagers are victims of disease like malaria, gastro-intestinal disease, skin disease etc which occur mostly due to usage of contaminated pond water. People do not get any alternative except to use water from one or two such ponds that exist in the villages. Well is the major source of drinking water. About 90 per cent of the households depend on such a source which is often in a very unhealthy condition and is the cause of water borne disease. Some cases of T.B. and leprosy are also registered in the study area. 258 individuals reported to be sick during the survey period. The sickness rate i.e., the number of sick per thousand population was about 303.17.

The crude death rate in our study area as recorded last year was 80.76 while infant mortality rate was around 107. This indicates the fact that the number of deaths of children below 1 year of age is more as compared to that of the general population. This is partly because they are not immunized and partly due to malnutrition.

Through our survey it is being revealed that majority of the women reported very poor availability of normal medical services. More than half of the women reported that they never register for antenatal care. Majority (78%) of births have taken place at home and deliveries have been conducted by untrained hands. The potential care have not been provided to most of the women (80%). More than 82 per cent of the married women are not aware about TT during pregnancy. They also do not have the correct knowledge about the age when the child should receive immunization. These points depict their ignorance in terms of health of women as well as children.

The low per capita income of the tribals fails to provide even one full meal a day for the members of the family. Thus the question of nutritious food for pregnant ladies is a far cry. The food consumed is not only insufficient in quantity but is unbalanced and low grade in terms of nutrient contents. This has a two-fold effect. On one hand it restricts the normal physical and mental growth of the people, especially the children, on the other, it is the cause of diseases like scabies, other skin infections, ricket, etc which mainly result from malnutrition and protein vitamin deficiency.

The tribal households suffer from acute housing problem. The various tribes like Lodha, Santal, Mahali, Shabar and Munda reside in mud huts with thatched roofs that are shabby and damp. A single room is shared by the whole family, sometimes by the goats and fowls too. Lodhas, in many places, do not have any proper shelter.

The agony of the rural mass is being aggravated, because they are still in the abyss of ignorance. The literacy level as depicted in Table 12.6 is not at all redeeming. The causes are poverty, lack of opportunity, lack of enthusiasm, lack of educational centres, distance etc. Women, in special, are seemed to be deprived of educational facilities. Both male and female literates have acquired some lower level education. Women are hardly registered in the relatively higher levels whereas men feature in these standards though their number is not very promising.

TABLE 12.6 : Literacy level of scheduled tribe population

Level	*Total*	*Male*	*Female*
Literate	43	26	17
Primary	59	33	26
Middle	44	29	15
Secondary	12	9	3
College	1	1	0
Illiterate	537 (77.16%)	277 (73.87%)	260 (80.99%)
Total	696 (100%)	375 (100%)	321 (100%)

Note: Children below 6 years of age are excluded from the total.

The existing facilities and amenities are not at all adequate. Out of seventy four villages have one-staff primary schools. There is one middle school in the entire region. Almost all villages have kutcha road links. Well is the only source of drinking water in all villages. But the wells are not at all maintained and often they are dried up.

As revealed from an analysis of the existing scenario, the entire set-up of our study area in terms of economic well-being, social condition, available facilities etc is in a deplorable state, which needs effective intervention and a plan has to be designed accordingly to meet the means to a sizeable yet feasible extent.

Microplanning for Integrated Tribal Development

The integrated development plan for tribals inhabiting seven villages of Jhargram Block would necessarily have two dimensions:

(a) the sectoral plans, and
(b) the spatial plans

The total plan would be designed based on the sectoral plans and spatial plans independently and their integration.

Sectoral Plan

The sectors of an economy could be divided into three broad categories.

(a) Sectors involved in production include
   1. Agriculture.
   2. Livestock and Animal Husbandry,
   3. Industries,
   4. Forest and Plantations, and
   5. Human Resource.

(b) Infrastructure which encompass sectors like.
   1. Road and Transport (Network facilities)
   2. Housing,
   3. Irrigation.
   4. Water Supply and Drainage, and
   5. Electricity.

(c) Service sectors include.
   1. Educational Facilities.
      School, Training Centres etc.
   2. Health and Welfare Services.
      Hospital/Health unit,
      Primary Health Care and Mother and Child Care Units, and
      Veterinary Centres
   3. Commercial Facilities.
      Markets and Marketing Channels.
   4. Other Amenities and Services.
      Library.
   5. Credit and Banking Facilities.
      Credit Societies, Banks etc.

Spatial Plan

The spatial plan would necessarily mean location identification of the new basic necessities inferred through the analysis and their representation on maps.

The Process of Plan Preparation

Every plan is derived within the broad framework of the aim of the study. In the present case the areas that are to be given prior emphasis are socio-economic upliftment of the tribals and poverty alleviation through suitable employment creation, subsequent income generation and provision of education, health facilities, etc.

The initial step in the process of plan design is basically an accounting of the demand and supply potential in terms of capacity utilization and generation. The dynamics of demand and supply is studied in details to bring out the loopholes and to identify areas where changes could be brought about or trace out new avenues. Hence, in our analysis, the prime demand constituent would be the unemployed or job seekers. The entire analysis has to be carried out in the individual sectors which should be followed by an integration of all the sectors to bring out a detailed encompassing plan for the study area.

The options for livelihood improvement and employment generation for the individual sectors have been described in the following sections.

Employment Generation and Livelihood Improvement

The age-sex distribution of the tribal population in our study area has been given in Table 12.7. This indicates the fact that about 65 per cent of the population are in the age group of 15-59 years, 30 per cent are in the age group of below 15 years and the rest constitute the above 60 year category.

TABLE 12.7 : Age-sex characteristics of scheduled tribe population

Category		*Persons*	*Percentage*
Total		851	100.00
Male		423	49.70
Female		428	50.30
Male	(Below 15 years)	136	—
Female	(Below 15 years)	125	—
Total	(Below 15 years)	261	30.67
Male	(15-59 years)	273	—
Female	(15-59 years)	279	—
Total	(15-59 years)	552	64.60
Male	(Above 60 years)	14	—
Female	(Above 60 years)	24	—
Total	(Above 60 years)	38	4.47

Disaggregating the population into gender classes, however, does not change the proportion significantly.

As observed in our study area the standard tribal household consists of 5 to 6 members of which the lead pair is involved in some economic activity and this pair is the bread earner for all its members. Apart form this, the extremely low economic conditions, absence of educational and related facilities and above all the lack of awareness among the people drive a section of the children to perform petty economic activities which are often of supporting nature.

Hence, the number of workers are 340 (40% of the total tribal population). Considering the fact that 65 per cent of the population are in the age category of 15-59 years, potential labour forces amounts to 552.

Therefore, the number of potential job seekers comes to 212 (552-348) or 25 per cent of total population.

Further it could be assumed that persons above 40-45 years have offsprings in the age group of 15-25 years who are already in the jobseekers' category. So, while planning for creation of job opportunities the section of the people in the 45-49 years age category who are still unemployed are not considered in the current planning exercise. Thus their number has to be subtracted from 210 i.e., potential jobseekers. On the other hand, considering the economic status of the population of the study area, a sizeable chunk of persons belonging to the class of workers may be earning meager amounts for which support from subsidiary activities turns out to be essential. Thus, the number of persons of such a class has to be supplemented to the job seekers category. Considering the above two aspects the number of potential job seekers amounts to around 200 and planning has to be directed towards the creation of employment opportunities for this section at present. It is expected that of the figure 200 around half will be women job seekers, so programmes should be designed to suit their capability and adaptability criteria. Again, considering the rate of population growth to be about 23 per cent, the number of potential jobseekers will increase by the same rate and by the end of our plan period (1994-1999) this number would increase to 225 and opportunities have to be created to accommodate the maximum possible number from this lot.

The Employment Plan

The existing structure of employment pattern as depicted in Table 12.8 indicates that agriculture provides employment to the majority of the workers. However, the concept of employment is always questionable as agricultural sector is characterized by underemployment and disguised unemployment to a large extent. Considering the condition of worker saturation thus agriculture does not leave much scope for activity generation. Rather a section of the agricultural lot could be shifted to other pursuits. Hence, policies and programmes could be designed for certain minor developments leading to livelihood improvement.

TABLE 12.8 : Distribution of workers under different activities

Category	*Percentage*
Agriculture & Agri allied activities	45
Agri labour	25
Forest	18
Rural industries	5
Other	7

However, the agriculture allied sectors which could be explored for activity generation potential sectors in this regard are livestock and animal husbandry, horticulture etc.

The forest sector should not be encouraged as a source of employment since this would automatically lead to destruction of forests. However, organised collection and marketing of forest produces, especially nontimber forest produces, could be a positive possibility. The major thrust sectors in relation to employment opportunity creation are rural industries and the service sector. Rural industries sector based on locally available natural resources opens up different avenues for employment. Again, need oriented services which are usually absent in remote rural areas, could be identified and suitable persons could be trained in such activities which would lead to opportunities for self employment as well as certain other attachments.

1. *Agriculture and Allied Sectors*

1A. *Agriculture*—Agriculture could be conceived as the back bone of the economy of the study area. A major portion of the population is engaged in agriculture, but often earning meager resources for their sustenance. Barring a few, most of the workers reside below the poverty line and are subjected to a bleak level of existence. Hence, increasing the income generation power of this activity turns out to be critical. The suggestions in this respect could be the following.

A1. Minor land improvement programmes mainly through introduction of irrigation facilities.

A2. Change in cropping pattern which would increase the productivity of land and generate income.

A3. Introduction of improved technology, seeds of HYV, and other inputs for increasing productivity.

A3.1 Training on the use of improved technology. (The prescription for improving irrigation facilities has been discussed in the section on Infrastructure)

Twenty-five per cent of land belonging to small farmers could be subjected to such improvements leading to increase in income. For large farmers the acceptance of the new methods is expected to be spontaneous and this would automatically improve their standard of living. The marginal farmers would be benefited to a negligible extent and avenues for subsidiary jobs for this section have to be chalked out.

1B. *Horticulture*—It is a new idea to be introduced in this region. Though people are aware of raising fruit bearing plants in their homestead lands it is not done in proper scientific way keeping an eye to a better production. Horticulture includes raising of vegetables, fruits, mushrooms, flowers, ornamental foliage and even aromatic and medicinal herbs. Thus it may generate some employment and thereby involvement in it as an occupation can add to one's income. Horticulture can be practiced in the following ways.

B1. Growing plants yielding vegetables, fruits etc in their homestead lands. This will include identification of suitable plants as well as cash crops, their growing and nurturing. Citrus fruits that grow well in this region, can be taken into consideration to start with.

B2. Growing of flower yielding plants like seasonal flowers which have market in the nearest cities like Calcutta or Jamshedpur.

B3. Nurturing of rosery can be done since the local soil is suitable for rose plants.

B4. Growing of mushrooms can be a very prospective kind of occupation. Looking at the demand of this vegetable and the simple ways for its growth, mushrooms can earn a good amount of cash.

B5. Herbs yielding medicinal extracts as well as aromatic substances can be grown which have good market value in drug industries and perfumery.

B6. Centres to be created for distributing saplings of improved varieties, mushroom spawns etc. for horticulturists and mushroom growers.

B7. Cooperatives to be formed to look after the sales and supply of the products to the market.

1C. *Livestock*—Livestock rearing is often carried out as an activity subsidiary to agriculture though the earnings are often meager. Hence, the improvement of the sector could be brought about through the following measures:

C1. Drastic cut down in the stock of nearly nonproductive cattle population which are reared mainly due to socio religious beliefs.

C2. Substitution of local variety by hybrid variety to increase the production of milk and cowdung. Cowdung can be used as biofertiliser and fuel by the rural people.

C3. Bullocks to be replaced by simple mechanical tillers wherever necessary.

C4. Introduction of the concept and creation of 2-cattle dairy 4-cattle dairy, 8-cattle dairy which would result in an organised collection and distribution of milk and would create additional job opportunities.

C5. Development of piggery, goatery, poultry etc. on a cooperative basis.

C6. Fish production is an activity which could be explored in the lateritic region.

1D. *Forest and Plantation*—In Jhargram block about 50 per cent of the land is covered by forests mainly owned by the Government. Inhabitants of villages proximous to the forest area earn their livelihood from forest based activities. But the erratic use of forest resources coupled with the process of natural degeneration in recent years has resulted in shortage of forest resources. Men often migrate to other regions in search of a source of income but women are left behind to face the drudgeries of life. Hence, the programmes related to forests should have a reasonable consideration for women. Such programmes could be of the following nature.

D1. Control unabated felling of trees and co-operate with the government in the programmes of forest management.

D2. Identification of non-timber forest produce (NTFP) which after semiprocessing and processing could be converted to marketable items like sal leaf plates, mats, seed beads, buttons, other decorative items etc.

D3. Identification and collection of indigenous medicinal and aromatic plants and their organised cultivation.

D4. Wasteland development through try techniques.
D5. Energy plantations in selected areas.
D6. Nursery raising on a co-operative or individual effort.

2. *Rural Industries and Crafts*

There is hardly any organised industry in the study area. The very few that exist are based on natural resources. The general economic condition obviously infers that the technology used is of a very primitive type leading to low levels of productivity and income. Hence, planning for the industrial sector will have two aspects—(a) setting up of new units, and (b) improving the productive capacity of the existing ones. The industrial units that can come up in this region are the following:

2A. Forest produce based units—this would include small units based on NTFP like:

A1. Sal leaf plate making
A2. Rope making with sabai grass
A3. Matmaking
A4. Oil extraction from seeds of eucalyptus, acacia etc.
A5. Making of ornamental and decorative items from seeds, dried fruits, etc.

2B. Preparation of extracts of medicines and perfumes from indigenous herbs and shrubs.

2C. Food processing units like:

C1. Rice husking mills.
C2. Jam, jelly, sauce making units etc.
C3. Pulse processing units.
C4. Cashew nut processing units etc.

2D. *Pottery*—The pottery industry could have two wings—one concentrating on the production of traditional earthenware and the other could be diversified in making decorative items to cater to the demand of urban markets.

2E. Printing and weaving industries.

2F. Bio-fertiliser units using local resources.

2G. Bio-gas plant.

2H. Service industries like masonry, electrical and electronic maintenance, etc.

3. *Infrastructure*

The infrastructural facilities in the area are of a very poor standard. So programmes should be chalked out in such a manner that infrastructure does not provide any hindrance to the general developmental programme for socio-economic uplift rather it should provide support in implementation. The measures for infrastructural development are mentioned below:

3A. *Irrigation*—A very insignificant portion of the cultivable land has irrigation facilities. But because poor irrigation often stands between the improved agricultural production minor irrigation projects have to be taken up on a large scale covering our study area. The projects could be of the following types:

A1. Creation of 'borobandh' in suitable areas.
A2. Excavation of dug-wells and tanks.
A3. Sinking of shallow tube-wells.

3B. *Provision of housing and sanitation facilities*—Shelter is one of the basic needs for any human being. As already discussed the state of housing and sanitation is deplorable and this is a primary cause of their ill health. The thatched mud houses, that too in a dilapidated condition characterize the shelter status. Often such houses do not have provisions for doors and windows. Programmes which could be taken up for improving the present stock and creation of liveable shelters for those who do not have are

B1. Improvement of the present housing.
  (i) Repair and replacement, wherever necessary, by using low cost housing material.
B2. Creation of few low cost houses.
B3. Introduction of better shelter concepts like, wide windows, pucca floor, separation of goatery, poultry etc.
B4. Provision of better sanitary facilities to improve the residential environment. Use of disinfectants in pits of stagnant water, marshy land, goatery, cowshed etc.
B5. Awareness campaigns for better living and better sanitation.

3C. *Transport*—Since our study area encompasses certain interior villages having no defined roads, the network plan should be such so that minimum accessibility is provided. Hence, the planning measure should be:

C1. Provision of kutcha link roads from all villages to their nearest service and activity centres.
C2. Repair, wherever necessary, of the existing approach roads.

4. *Provision of Basic Health Facilities and Nutrition*

The existing scenario portrays multiple problems in terms of health, nutrition, education etc. The utter backwardness in their physical and mental conditions thus requires a special attention and programmes have to be chalked out accordingly. The Table 12.9 on the age-sex and marital composition of the ST population gives an indication that specific provisions are to be arranged for women and children at specific ages.

TABLE 12.9: Population by age, sex, and marital composition

		Marital status				
Age	*Sex*	*Single*	*Married*	*Widow*	*Divorcee/ separated*	*Total*
(1)	(2)	(3)	(4)	(5)	(6)	(7)
0—4	M	53	—	—	—	53
	F	48	—	—	—	48
	T	101	—	—	—	101
5—9	M	53	—	—	—	53
	F	55	—	—	—	55
	T	108	—	—	—	108
10—14	M	30	—	—	—	30
	F	21	1	—	—	22
	T	51	1	—	—	52
15—19	M	39	3	—	—	42
	F	34	26	—	—	60
	T	73	29	—	—	102
20—24	M	20	44	—	—	64
	F	17	37	—	1	55
	T	37	81	—	1	119
25—29	M	8	45	—	—	54
	F	9	48	1	1	58
	T	17	93	1	1	112
30—34	M	2	57	—	—	59
	F	1	50	—	—	51
	T	3	107	—	—	110
35—39	M	1	30	—	—	31
	F	—	25	1	—	26
	T	1	55	1	—	57

(*Contd.*)

TABLE 12.9 (*Contd.*)

(1)	(2)	(3)	(4)	(5)	(6)	(7)
40—44	M	1	1	—	—	2
	F	—	2	2	1	5
	T	1	3	2	1	7
45—49	M	—	2	—	—	2
	F	—	1	3	1	5
	T	—	3	3	1	7
50—54	M	1	9	—	—	10
	F	—	6	4	—	10
	T	1	15	4	—	20
55—59	M	—	10	—	—	10
	F	—	5	3	—	8
	T	—	15	3	—	18
60+	M	—	14	—	—	14
	F	—	14	8	2	24
	T	—	28	8	2	38
Total	M	208	215	0	0	423
	F	185	215	22	6	428
	T	393	430	22	6	851

It has been observed in our study area that the average age at marriage is around 16 years for women and age at first delivery is 18 years 6 months on an average. Moreover, the crude birth rate is 32.7 which means that in one year considering the total households 26 live berths have taken place. The age specific marital fertility rate is shown in Table 12.10.

TABLE 12.10 : Age-specific marital fertility rate

Age-group	*PMW (presently married women)*	*No. of live births*	*Age specific marital fertility rate*
20—24	37	10	270.27
25—29	48	11	229.16
30—34	50	4	80.00
35—39	25	1	40.00
Total	160	26	162.50

It can be observed that the marital fertility rate of women is highest in the age group 20 to 24 years followed by that of 25 to 29 years.

Thus, considering the facts that 26 children are born in a year nearabout 25 mothers have to be cared in their prenatal and postnatal stages. Again out of these 25 others about 50 per cent have to be provided with special facilities as they are likely to deliver their first baby.

Keeping in view the features of the health and nutrition condition the programmes that could be sorted out are

A. Provision of antenatal and postnatal care in terms of:

I. *Prenatal care*

(i) Vaccines for women in the antenatal stage.
(ii) Medicines like iron, calcium and multi vitamin drugs.
(iii) Nutritious food comprised of rice, lentils, green vegetables, and/or soyabean products.
(iv) Free medical care periodically

II. *Postnatal care*

(i) Nutritious food
(ii) Free medical care if required.

B. Improvement of nutritional and health status of children.

(i) Provision of full meal for children
(ii) Provision of nutritious food as mentioned above
(iii) Provision of baby food, tinned milk etc for infants below one year of age.
(iv) Immunization of all infants below 2 years of age.

C. Provision of centres for preventive and promotive work for MCH, immunization, family planning and prevention of water borne diseases.

(i) Establishing contacts with the Government health centres.
(ii) Provision of outreach health care facilities through mobile clinical services.
(iii) Provision of medical and paramedical staff for carrying out the services. Medical staff could be hired from outside while paramedical staff and assistants could be trained over the plan period by selecting suitable persons from our target population.

D. Health awareness campaigns and training on an extensive scale among people especially women in the line of personal hygienes, child care, antenatal and prenatal care etc.

E. Considering their preference for wells as against tubewells, provision of wells in each and every village and their proper maintenance by using disinfectants and carrying out the chlorination operation.

F. Collaboration with existing voluntary agencies and incorporation of the task of monitoring and the follow-up on them.

G. Provision of facilities for personnel of voluntary agencies who would carry out the follow up action.

5. *Provision of Educational Facilities and General Awareness Creation through Formal, Non-formal or Adult Education Programme*

Like all other sectors of socio-economic life, educationally the tribal people are at a very low level of development. On the whole, formal education has made very little impact on tribal groups. The level of education has already been depicted by the table on literacy rates in the previous section. As evident, the literacy rate in this area is around 20 per cent which is definitely very low. So efforts have to be made in this direction as the level of education and educational awareness act as vehicles for any socio-cultural and economic uplift.

The primary target group in the process of educating the people would be the children below 15 years of age. Educating adults should also be taken up as a subsidiary objective but prime importance should be assigned on efforts for awareness creation among the adult male and female section of the population.

The Table 12.11 gives an estimate of the facilities to be created for the children under different age groups.

TABLE 12.11 : Breakup of children below fourteen years of age

Age group	*No. of children*
< 3	66
3 — 6	57
7 — 14	138

Programmes that have been chalked out are listed below:

A. Provision of creches and day-care centres for the infants (below 3 years) to be created so that they would be brought up in a more congenial surrounding when both parents are out for work. This would serve the dual purpose of bringing up and carrying the infants by a local lady baby sitter (who could be trained for the job) while their habit of attending creche would help them in going for school education. Such facilities should be located at commutable distances from each village. Hence considering the study area around 1 to 2 creches or day-care centres could be suggested.

B. Provision of non-formal education for children in the age category 3-6 years.

(i) Provisions of set-up which could be in terms of a shed, a room, a community place etc.
(ii) Incentives to children for attending such schools. These could be in the form of midday meal, books, clothes or cultural programmes.

Such schools could be located in each and every village or one for every 2 to 3 villages depending on the population size of the villages concerned.

C. Provision of primary and secondary education:

(i) Revival and improvement of conditions of existing primary schools in that area.
(ii) Creation of one-teacher primary level schools where no provision exist.
(iii) Provision of incentives in the form of a principal meal, uniforms, scholarships for achieving specified standards.
(iv) Establishing contacts with secondary schools and provision of incentives for attending such schools.

D. Creation of night schools and centres for carrying out adult education programmes.

E. Trainings, workshops and campaigns on an extensive scale to be carried out to create an awareness and interest in education.

F. Periodic monitoring by voluntary organisation of the actual happenings in this sector after the programmes have been implemented and provision of facilities for personnel who would be endowed with this task.

Concluding Remarks

This microplanning exercise is based on the primary aim of socio-economic upliftment. The analysis of problems in every sector and its remedial measures have been in consonance with the people's perception and knowledge of the perspectives. Adequate income generation and resulting economic betterment utilizing one's indigenous resources sets in an overall multiplier effect in terms of improvement in the quality of life. Improvement of life implies better living conditions, greater awareness and consciousness and an improved effort for getting involved in the continuing development process. This participatory approach for plan formulation at the micro level would automatically induce a sense of belongingness among the people which is the key to success of the sustainability of any programme.

REFERENCES

Belshaw, C.S., 1972. 'Development : The Contribution of Anthropology' in *International Social Science Journal*, Vol. 24, pp. 83-94.

Food and Agriculture Organisation of the United Nations, 1987. Training for Decentralised Planing, Rome.

FAO, 1985. Towards Multilevel Planning for Agriculture and Rural Development in Asia and the Pacific, Paper No. 52.

Mathur, O.P, 1981. Training for Regional Development Planning, Perspectives for the Third Development Decade, UNCRD, Nagoya.

Ministry of Home Affairs, 1977. Approach to tribal development in the Sixth Plan: A preliminary perspective' in Occasional papers on tribal development, No. 17, Govt. of India, New Delhi, pp. 98.

Narayan, S., 1984. Plan and Planned Development—Case Study of a Primitive Tribe, *Mainstream*, Vol. XXII, No. 49, August.

Report of the working group on development of Scheduled tribes during Seventh Five-Year Plan, 1985-90, Government of India, MOHA, New Delhi, December, 1984.

Vidhyarthi, L.P., 1981. *Tribal Development and Its Administration*, Concept Publishing Co., New Delhi.

Vidhyarthi, L.P. and B.N. Sahay, 1980. Applied Anthropology and Development in India, New Delhi.

13

SUSTAINABLE AGRICULTURAL DEVELOPMENT

B.L. SHARMA and PALAK BHARDWAJ

Introduction

MAN'S efforts in waking human civilization has been a strenuous one. It has undergone a series of historical and social changes through ages. The present stage of development of human civilization has thus, reached upon a successional and sequential changes one over the other. The first imprints of the man over this earth as a cultural means, appeared long back with the introduction of the cultivation of plants, which settled him from a plundering to sedentary life as back as about 10,000 B.C. now. Thus from food gatherer man had graduated to food producer. Since then there was a tendency for human being to concentrate in the places, where agriculture is feasible with least hurdle.

At this stage, the dispersal of population distribution, growth and density were in accordance with the environmental suitability for agriculture. Therefore, the higher concentration of population became the hubbub of major activities with agriculture as a predominant occupation. China, India, South-East Asia and coastal regions of Asian continent and Southern Europe accounting for about two-third of the world's total population, became the main centres of the activities. Following industrial revolution, the western countries modernised their agricultural practices, resulting in high productivity and thereby environmental factors are greatly controlled and modified. However, in the tropics the agriculture remained much akin to the ecosystem with little progress in production front. It is only recently that modern inputs are increasing by being used for agriculture in the tropical countries like India.

No doubt, we have been able to harvest more production from the fields with the aid of HYV seeds, chemical fertilizers, and large scale mechanisation of farming. Further, use of pesticides, spray powders and chemical compounds have helped greatly in increasing food production and solving our food problems. But is the accelerating pace of agricultural development really viable? Are we not derooting the genesis of the plant breeds changing in their ecosystem? These are the questions of the present day relevant particularly in the context of sustainable agriculture.

Concept of Sustainable Agriculture

In under-developed countries like India, the existing level of development does not rise upto the expectations of people as a large section lives below poverty line.

A sustainable agricultural system in India will be such that it will ensure sufficient production to *meet* the social *demand* in one hand, on the other hand, the ecosystem must not be disturbed beyond its self ameliorating capacity, so that future generation can also meet their needs.

India has witnessed three successive stages of agricultural development during the last forty years. They are:

1. Extension of cultivated land,
2. Intensification of agriculture, and
3. Modern techniques of bioculture etc.

Because, any level of agricultural development in all the three stages was not sufficient to cope with the increasing population in the country. The steady growth of population in almost all the countries of the world has compelled the society to harness more and more yield from the finite agricultural land resulted in retardation of land and degradation of agroecological balance of the nature. The persistence pressure of population on agricultural land changes the whole dimension and complexion of the agricultural development. Grigg (1976) has rightly observed the situation and said that "Population growth is a function of change in agriculture".

Plant and Ecosystem

The basic assumption of change in agriculture is based on the fundamental principles of the plant and ecosystem. Because, every plant and animal sustain life upon an exact ecosystem in which they thrive. Thus, the basic principle of growth and development is related with the certain laws of nature such as:

1. Tolerance capacity of plants (species),
2. Isolation,
3. Growth development, and
4. Edaphic conditions.

The tolerance capacity of plant refers to the ability of plant to withstand in a varied conditions of environment, whereas the isolation speaks of the regional or areal demarcation which differentiates one plant community (species) from another, that means the occurrence of plant in different territory, or region is a real identification of plant community or species that distinguishes each other. Thus plant community may be described as a similar structure or morphology of plant species.

In an ecosystem, only those plant species can sustain which have adaptability with the existing environment and hence, differ from place to place according to the change in ecosystem. This speaks of the diversity of plant and animal life on this earth. Thus, the growth and development is a factor of adaptable ecosystem which involves the edaphic conditions in which varied types of plant germinates and sustain upon it on a favourable environment. The varying number of ecological races or ecotypes of plants is nothing but a plant association in which varied plants exist simultaneously or withstand in one ecotypes (Sharma and Bhardwaj, 1993).

Looking to the above fundamental aspects of agroecology, the question of sustainable or compatible agriculture is a crucial issue in development, which has a long historic past as traced earlier in the text.

The Present Scenario of the Land Potentiality

The most revealing factor of our agricultural economy today, is rather shocking one. If we look to the figures of our land resources and the degradation of land in a gradual process of competing ecosystem, we have about 143 million hectares of agricultural land which count nearly 43 per cent of the total mass area of 329 million hectares, of which about 40 to 60 million or 29 to 42 per cent land is already degraded. Our forest area, out of the 75 million ha. about 40 million ha. is considered to be degraded in the form of exposure (without tree cover) and scanty shrub cover. About 50 million ha. land is taken as unproductive while, looking to the present need of foodgrains, we needed additional 10 million ha. of land to cope with the increasing population by the end of the century.

The loss of arable land by natural hazards is no less significant when we look to the fact that about 1 million ha. of land is affected due to shifting cultivation every year and about 8000 ha. is encroached by ravines and gully erosion. About 8.4 million tonnes of nutrients are washed away from the topsoil annually, while about 1.5 million ha. of forests are reported to be lost every year.

In fact, about 70 per cent of our arable land is rain fed. The unirrigated dry lands accounting 45 per cent of total cropped area contributes 42 per cent of the total output of foodgrains.

Therefore, it is imperative that we will have to put more efforts to sustain upon the existing land resources. But this fact should also be borne in mind that whatever land at present is available for cultivation is not enough to support the increasing population if the land degradation process is allowed to go on at the present rate.

In this context, it may be added here that human efforts in regaining the vitality of the degraded land are also on the rising trends. India has made remarkable progress in developing irrigation, fertilizers, and developing various techniques of bioculture and genetic engineering in the recent past.

Sustainability of Agricultural Development

The analysis of the sustainable agricultural development may be perceived and viewed from the case studies carried out in Rajasthan State and achievements reached here in recent past.

To study the measures of sustainable agricultural development in the state, an analytical approach is adopted to encompass the development in different sectors of the technological applications i.e., use of chemical fertilizers (NPK), use of HYV seeds and raising productivity, efficiency and fertility of soils, etc. This is in order to locate whether, we have been able to achieve the goal of sustainable development through our efforts in this direction or merely to achieve temporal justification.

Let us first, examine the case of the use of chemical fertilizers (NPK) and the consequences reached thereupon.

Case 1: Statement—The Use of Chemical Fertilizers (NPK)

The introduction of chemical fertilizers (NPK) in India has not been very old, but a recent phenomenon. The launching of Green Revolution (GR) (1968), coincided with the widespread use of chemical fertilizers in India. Though, its use before GR has been a very scarce and limited, the period marked as GR, has opened the doors of the farmers in most of the states like Panjab, Haryana, Western UP, Andhra Pradesh, Maharashtra, Tamil Nadu, Orissa and Rajasthan etc. Thus, the susceptible attitude of the farmers was replaced by an open mind who in turn used Chemical Fertilizers (NPK) without any consideration. This led to the revolution in agricultural sector and resulted in bumper production of crops (foodgrains) under the government supervision at many places. The self-sufficiency in production of foodgrain was achieved. In recent past, the target was marked above the expectations i.e., 150 million tonnes in 1989, 170 million tonnes in 1991 and 180 m tonnes in 1993. Though all of it (high production) was not due to the use of fertilizers alone, HYV seeds, and modern technology had also contributed.

While analysing these factors, we may ascertain the factors attributed to change and effectiveness of the inputs. The method of analysing effectiveness of the factor, is known as 'Factor Analysis' using Monte-Carlo simulation method widely used in economics.

Monte Carlo simulation model determines the value of controlled variables at testing of the effect of the changes in probability value and parameters (Collenutt, 1972).

The application of the Monte Carlo method, indicating the rate of return of the value input/output, determines the loss or gain of the used item. We may determine whether it is useful or not. Thus, the probable value is known for its future application. This technique (Monte Carlo), was used to evaluate the probable rate of return and the probabilities of getting return in some specified range (Meyer, 1956). Thus, it helps in knowing the future projections of the use or misuse.

B. *Hypothesis*

Whether the incremental use of chemical fertilizers (NPK) is useful in foodgrain production.

C. *Methodology*

Owing to the lengthy procedure of Monte Carlo technique, it is difficult to make use of this technique. Therefore, a simple result deducting procedure is adopted for the present problem of testing the hypothesis whether use of Fertilizers (NPK) is useful or not in the present state of affairs. In the beginning, there was a hue and cry over the use of fertilizers, but gradually the demand has been gone down as it appears that farmers are not more interested in its use. For the, simple reason, a test has been carried out on the basis of input/output ratio, on some different crops i.e., wheat, bajra mustard etc., that how and to what extent the fertilizers as an input, factor, has been useful in yielding proportionately to production/output.

D. *Explanation*

Generally, the use of chemical fertilizers is made according to the soil properties. It is suggested that NPK should be used in the ratio of 4:2:1 but, it too, depends on the soil quality hence, it differs from state to state and place to place. Inspite of all, nitrogen is used extensively.

The regional disparity of fertilizers' use and imbalances are seen from region to region. In 1979, southern India was leading in fertilizers consumption but, later, northern India particularly, Punjab, topped the list with consumption 149 kg. per ha. followed by Uttar Pradesh. Whereas Andhra Pradesh, Maharashtra and Bihar, consumed about 11.8 per cent, 7.6 per cent and 8.3 per cent respectively, the consumption in Uttar Pradesh is the highest by 21.9 per cent and in Punjab it is 12.8 per cent, in 1984-85. (*Yojana*, November 1987).

The consumption of NPK was higher in irrigated areas as recorded by 90 kg or even 128 kg per ha while rain fed areas consumed only about 2 kg per ha.

In 1992-93, on all India basis, the average National Yield of some of the foodgrains was totally different in areas of their cultivation and in farms operated by agricultural department, as listed in Table 13.1.

TABLE 13.1 : Comparative statement of yield rates

Crop	*National average Yield Qtl./per ha.*	*Agrl. farms Qtl./per ha.*
Wheat	14.10	38.14
Paddy	12.35	31.49
Jowar	5.91	35.54
Maize	12.93	32.28

Source: ICAR Bulletin, 1984.

From the Table 13.1 it may be drawn out that in government operated farms, the yield is much more than the ordinary fields. It is because of the intense care and meeting all other necessary requirements of the crops. But, it may also be noted that if proper care is taken and meet out by farmers, they can get better results. The basic requirements are: the proper ratio of chemical fertilizers, adequate water, and High Yielding Variety seeds. The soils must have been tested for the proper treatment and use of NPK. And above all capital should be available for inputs in hand.

The question is whether an ordinary farmer may provide all these basic needs or can avail the opportunity of using all that?

The cost of fertilizers is much more in India as compared to other countries viz., one kg. of supper phosphate costs 8.9 kg. of paddy in Pakistan, 1.92 kg USA, 1 kg. in Japan and 3.19 kg. of paddy in India.

The consumption and decennial growth of chemical fertilizers in the districts of Rajasthan speak of the overall high increment one year after another (Table 13.2), but if we compare the input/output ratio then the emergent picture is quite different from our expectations.

TABLE 13.2 : Use of chemical (NPK) fertilizers in the State

(*In M. Tonnes*)

Data	*1960-61*	*1970-71*	*1980-81*	*1985*
Ajmer	25	482	1876	3888
Alwar	23	4326	6149	—
Banswara	12	636	1891	—
Barmer	11	—	5	447
Bharatpur	25	2290	5323	—
Bhilwara	—	2541	4536	—
Bikaner	—	—	1337	5322
Bundi	58	2228	9242	—
Chittorgarh	—	5115	10745	—
Churu	18	—	115	—
Dungarpur	11	141	568	—
Ganganagar	386	13693	41898	55162
Jaipur	29	2303	8148	—
Jaisalmer	—	—	—	—
Alore & Barmer	16	553	2352	2747
Jhalawar	—	18	968	1595
JJN & Sikar	19	—	—	11965
Dholpur & Jaisalmer	6	997	1396	2747
Kota	36	10178	15812	—
Nagaur	28	175	1137	—
Pali	68	2074	4287	6824
S. Madhopur	210	1098	7624	—
Sikar	7	683	2529	1979
Sirohi	15	339	1408	—
Tonk	59	317	1574	—
Udaipur	89	2693	3681	—
Total	1168	53721	135104	—

The Table 13.2 reveals the fact about the consumption of fertilizers and production trends during the last three decades in the state. If 1961 is taken as the base year equal to the index of 100, then the deviation of fertilizers growth recorded is 4.5 times in 1970-71 period and 11.6 times rise in 1980-81 period. But, if the growth of foodgrain production is recorded it comes to 1.9 and 2.3 times rise over these periods respectively. If the input/output contrast between foodgrain as output and fertilizers as input is drawn, then the picture is emerged as 4.5 input ratio to 1.9 output index in 1971, while in 1981 it comes to 11.6 to 2.3 index, respectively. Thus a difference is marked as 2.6 and 9.3 index value over 1961 respectively. This trend clearly provides the explanation to the fact that the input value is much more in comparison to the output value. It also may be born out that the fertilizers input is not very much encouraging as it has been envisaged by the experts. The findings of the Agricultural Extension Research Groups in the state, may be summarized in the following way:

In the year 1987, except Bajra, percentage of farmers using fertilizers has increased by 14 to 41 per cent in different Kharif crops. Although, consumption of fertilizers per hectare has decreased from the normal period of 1987 season as it was the worst so far the monsoon is concerned, even then percentage of fields being fertilized has increased which shows that there is a great impact on the contact farmers and extension agencies.

Case 2: Statement of the Problem : HYV Seeds

The wide ranging application of High Yielding Variety seeds (HYV) has hasten the boosting of production of foodgrain crops in 'patents' on international level through genetic engineering. This has widen the course of variety of plant cultivation but has tighten the limits of their growth in a particular manner. Because, the genetically developed (Patent) seeds, would be checked through patent-brand of pesticides, which would be available only with the limited authorized dealers (International Agents). Thus, the international control over these patents would be monopolised by a limited nations, and hence, will control the whole system.

B. *Hypothesis*

High Yielding Variety seeds have reduced the risk of foodgrain production and thus have helped in increasing the foodgrain in India in general and state in particular.

C. *Explanations*

During this period, if we take the use of fertilizers into account then we find that there has been an incremental trend throughout the period. The state consumed about 460.8 metric tonnes of nitrogen, 9,403 metric tonnes (mt) of phosphorus and 3,135 mt of potash in 1977, which rose to 1,06,623 mt of nitrogen, 2,298 mt of phosphorus and 5,499 mt. of potash in 1981, indicating the overall rising trend in contrast to it.

The use of High Yielding Variety (HYV) seeds in 1971 and 1988 to net area sown is an another attributable landmark in making agriculture a viable economy in the state. The area under cereal crops of HYV seeds has been increased in almost all the districts with a high level technological inputs of various aids in order to achieve better results. Despite the fact that scientific agriculture is being practiced in most of the areas, the productive index available to this date, indicates the decreasing trends, reflecting our inability to achieve what we expected to this date.

These two very factors i.e., uncertainty, and modernity, led our state of agriculture in a variety of crops, on a fractional part of the field with long fallow system to rotation and, inter cropping to mixed cropping systems, site selection in accordance with the water, temperature, and precipitation makes an important factor together with modern inputs as a visible considerations to maintain balance between growth and production.

Case 3 : New Techniques and Technology

In the span of agricultural development, plant protection techniques have become a major part of the field operations. Because, many ameliorating efforts and techniques have developed immunity in many pests, termites, and insects as well as provided the favourable conditions for the congestion of harmful bacterial growth. The increasing crop diseases and losses of production are the evidences in support of the thesis. Thus, the side effects of plant protection technique cannot be over emphasized and looked upon as only the means of the future development. In the wake of plant protection, other parallel techniques have been developed. These may be termed as the developing span in agriculture.

1. Genetic engineering,
2. Modern methods of synthesis,

3. Energy farming techniques,
4. Data processing method, and
5. System concepts, etc., etc.

As it has already been discussed in the earlier pages that the future of chemistry closely linked with the agricultural development, sets the goal of developing trends as mentioned above.

The genetic engineering is a recent development in plant growth and development management system. It indicates the plant breeding system. Method involves in it is D.N.A. transfer system with agro-bacterium and direct gene transfer system. Genetic engineering is developed in order to improve quality to raise resistance to diseases, and to increase herbicide, resistance power.

All these measures in modern agriculture, are considered a boon to the agricultural development. And, there is no doubt about it, that agriculture has made unprecedented rise over the past years. Thus it may be said that it is a good start, but in fact, is it sustainable?

The outcoming results and the consequences of the modern inputs, are no way simple ones. The reactions and the reflections of all these modern technologies, are some way shocking one, and production value in some cases is decreasing drastically. If the 'Patents' are not used, we may not get even a single grain from the fields, which used to provide us ample food.

So, the whole operation needs a thorough diagnosis not the patent dozes of all kinds to make experiments one after another.

Conclusion

Looking to all these facts we may come to the conclusion that the present attempt may not be considered a final job in highlighting the problems. It may be proved to be a significant attempt on the problem of agro-ecology, directly now cropping up with the developmental process and the reactions which are yet to be seen ahead. In this context, the present study sheds light on the problem areas and their treatment for the new cropping patterns in time to come. It may be considered as the basis for the global problem.

This problem is almost identical in all the developing countries, where inputs of compatible means are common in all the traditional bound agricultural practices. India provides a best example of such problems with which it is confronting today.

In a sustainable agricultural development, with reference to the present context we have to see the recent developments like bioculture, genetic engineering, and agro ecological problems more particularly in traditional countries like India, in the light of environmental compatibility of agriculture with Risk Spreading agriculture, future of chemistry in agriculture, and new technology etc.

REFERENCES

Collinutt, 1972. *Progress in Geography*, Vol.2, p. 139.

Meyer, H.A., 1956. Symposium on Monte Carlo Methods.

Sharma, B.L. and Palak Bhardwaj, 1993. *Agro Ecological Problems*, Daya Publication, New Delhi.

Yojana, 1987. November issue.

14

FARMING SYSTEMS RESEARCH AND SUSTAINABLE AGRICULTURE
Lessons from a Field Experiment in West Bengal, India

S.S. GHATAK

INTRODUCTION

THE history of agricultural development in India, with special reference to green revolution revealed that the technologies were primarily designed for the favoured classes (resource-rich) of farmers and for the best endowed agro-climatic regions. Consequently, these technologies, in most cases, spread slowly to the poor farmers (resource-poor). Besides, the technologies also appeared to have a number of short to medium term problems e.g. increasing pest, disease and weed problems, deterioration of soil fertility and dependency on mineral fertilizers etc. One had to tackle these various issues individually as it appeared. But at the present day situation there has been a growing realization that many, if not all, of the problems, are essentially systemic in nature (Conway, 1985). They are linked to each other and it is better that one should handle them holistically.

The earlier approaches of the scientists to find out the solutions of the problems were mostly influenced through their biases and more particularly through their discipline embedded views which are now-a-days referred to as "Reflection Lens". These often resulted to mismanagement in setting up an equilibrium between 'demand pull' and 'supply push'. The research approaches have always been to "reduce problems into manageable components" (Fig. 14.1).

On the other hand, the Farming Systems Research and Extension (FSR/E) approach is more holistic and systematic in nature (Fig. 14.2).

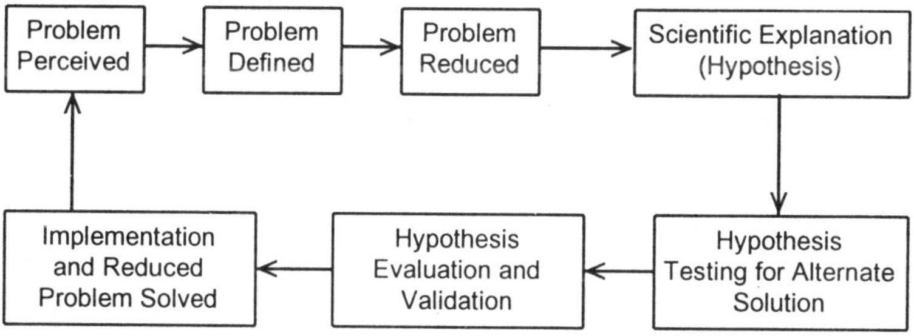

FIG. 14.1. Conventional approach: quasi solution

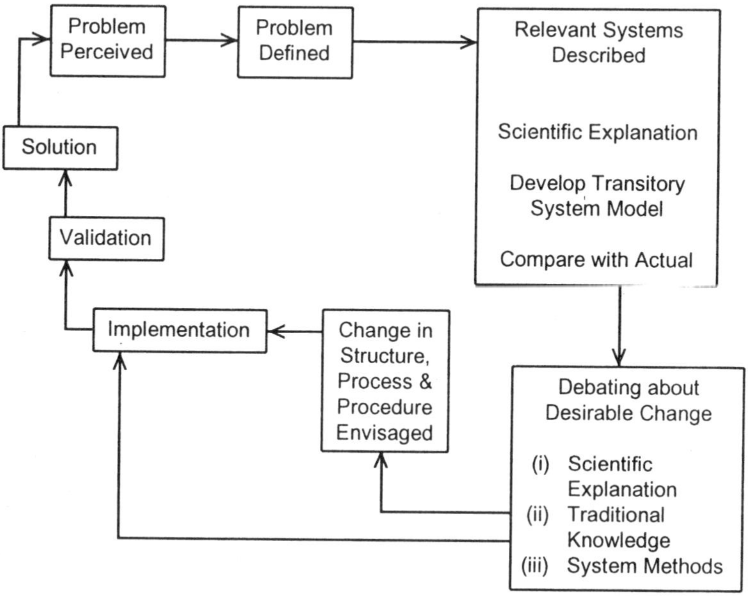

FIG. 14.2 : FSR/E Approach: system concerned

In a broader sense, the main objectives of Farming Systems Research and Extension (FSR/E) approach are to augment income and develop technologies for sustainability of income through proper use of locally available resources which should be economically feasible and environmentally viable with the ultimate aim of improving the quality of life. While considering agricultural production, one should keep an eye to the other attributes of production e.g., productivity, stability, sustainability and equitability.

Sustainable agriculture is a fascinating subject. Its romance lies,in part, with broad appeal to agriculturists. Numerous definitions and interpretations of sustainable agriculture have been in vogue, even though most agriculturists, nonetheless, have their own favoured version. However, many of them claim that they alone have 'the answer'. While practicing FSR, the researchers must not adopt this habit which fails to recognise that the contributions of several disciplines are necessary to foster sustainability in agriculture. Everybody must contribute and everybody has a role (Lightfoot *et al.*, 1993).

Materials and Methods

I.A. *Exploration of Priority Problems*

Each problem was prioritized in sequence. The first step in this exploration was to determine whether the problem was adequately defined e.g., poor crop stand establishment may have been observed during the initial diagnosis. It was critically analysed as to whether the poor stand counts were due to variations in timing of planting, expectations of high or low moisture regime or use of excessive fertilizer, poor quality seed, wrong depth of planting, pest attack on the seeds after sowing or after germination etc. Each cause implied a different thrust and hence, a different experimental content. This multiplied (and not reduced) the points of attack on the problem and increased the probability of finding an acceptable solution. In looking for options for solutions the FSR/E team tapped all available sources of information, national and international literatures, their own knowledge of biology of the problem and its causes particularly the experience of relevant research specialists. Specification of the best bet (Collinson, 1987) under present knowledge allowed specification of the level and timing of labour and cash requirements for the adoption of the solution by the target group farmers. If technical solution was not possible from the past results then a clear statement of the problem and a description of the possible approaches to its solution were passed to the appropriate specialist as agenda for technical 'On station' trial. Alternatively, specification was achieved from the results of exploratory or determinative experiments implemented as a part of the 'on farm' research programme. The whole system thus worked on the philosophy—'farmers first and the last (Chambers, 1986)'

I.B. *Target Grouping*

Target grouping is a stratification device which creates a framework for the research and development responsive to its clients, the farmers, who take the final decisions on technology adoption (Collinson, 1987). Following this each target group was composed of farmers operating in the same system in fairly homogeneous local circumstances, who were expected to have the same problems and opportunities as also who were expected to take up the same new technology. Where minor but important variations in local circumstances occurred, the target groups were further divided into 'recommendation domains' and eventually a specific recommendation evolved out through FSR/E process for each domain. The methodologies for identifying farmers' systems problems analysis leading to what problem to explore and what solution to test were adopted following Lightfoot *et al.* (1988). The methods developed first identified farmers systems—problems, second, analysed a key problem and third elaborated experimental hypotheses. The farmers participated throughout in the processes.

(i) *Identifying Systems—Problems*

The farmers were assisted to list issues they talk about, to select one issue for further exploration and to decide which systems-problems they want resolved. Group meetings with researchers and the farmers listed subjects of conversation. Asking farmers what they talk about stimulated more dialogue and proved a better entry point than asking direct questions about problems. During the process some farmers suggested a visit to their fields as adequate explanations could not be given in the office. Issues were elaborated further in one or two hour discussions on the farms of key informants. These discussions produced several problems for further study. However, in another group meeting with the same farmers obtained consensus on which problem should be studied first.

(ii) *Analysing Systems—Problems*

Systems—problems were analysed using the systems diagram based on those of Spedding (1979). Guide topics are framed in collaboration with 5 farmers and were administered to 20 randomly selected households from a total of 100. Interactions between farmers and researchers were more like informal conversations than formal interviews. Survey response provided information on bio-physical causes and socio-economic constraints surrounding the problems. Drawn on a black board, each cause and constraint was ascribed a box. Five 'Key informants' then explained the relationship between boxes and problems. Arrows were drawn to show these relationships. From the boxes and arrows a circular 'systems diagram' was constructed with bio-physical causes occupying one side and socio-economic constraints the other (Fig. 14.3). Each box formed one segment in the concentric rings surrounding the central problem. The segment size equalled the proportion of farmers who responded to that point against all points in that ring. All respondents were given an opportunity to comment on the diagram before it was finalised.

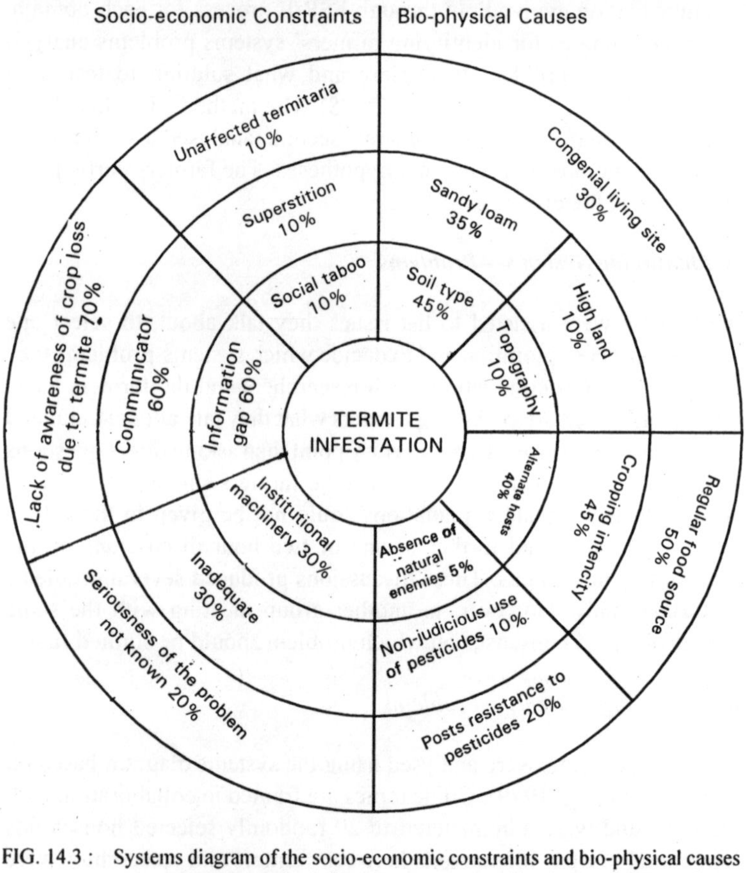

FIG. 14.3 : Systems diagram of the socio-economic constraints and bio-physical causes concerning the termite infestation as perceived by farmers of Baganipara, West Bengal, India

(iii) *Elaborating Farmers Experiments*

Experiments were elaborated by searching for potential solutions and screening out one that was worth trying and defining hypothesis for testing. The potential solutions were sought using the system diagram. The outer circle of the bio-physical causes indicated where interventions could be explored to solve the central problem. Key informants and researchers presented their ideas and the group debated the advantages and disadvantages. Experimental hypotheses were defined to address unanswered questions. The purpose of the experiment was to find answers to what the farmers did not know. After conducting the experiment it was concluded that participatory methods that use farmers' knowledge and systems logic are now solving problems that conventional cropping pattern research was incapable of addressing.

I.C. Ranking of Problems

Although it may be easy to describe many problems encountered in a given area but it becomes usually impossible to investigate more than a few at a time and as such priorities must be set accordingly. An initial ranking helps to establish a sense of priorities. Problems that receive a low ranking but having important relations to other problems or which have easily accessible solutions may certainly be addressed in the experimental programme planned. Problems that receive low ranking in this review and lack either obvious interactions with other problems or easy solutions may be eliminated from considerations altogether.

However, the problems should be ranked every year because with lapse of time the importance of certain problems may diminish. But the contents of 'on-farm' experiments should exhibit a logical progression from year to year, rather skip from one topic to another. The methods suggested were based on three criteria:

- — distribution of the problem.
- — importance of a particular crop enterprise to the farming systems of the target group.
- — loss of yield or income for which the problem is responsible.

Using the philosophy and methodology of FSR/E, an 'on farm' trial was conducted to find out alternative approaches to termite pest management at Baganipara area of Chandirampur village in Nadia district, West Bengal. Termites cause considerable damage not only to crops grown but also to the dwelling houses in this area. Due to superstition the farmers dare not to destroy the termitaria at least in their residence. Thus, considering the problem—casual linkages (Fig. 14.4) on termite infestation, as emerged through threadbare discussion with the farmers of the area, a hypothesis was formulated as "Crop replacement as one of the measures to escape damage from termites" in the cultivated land. The hypothesis formulated was supposed to satisfy the following:

— to make a break in the regular food supply of termites to create bigger harvest from a sustainable low-polluting agriculture

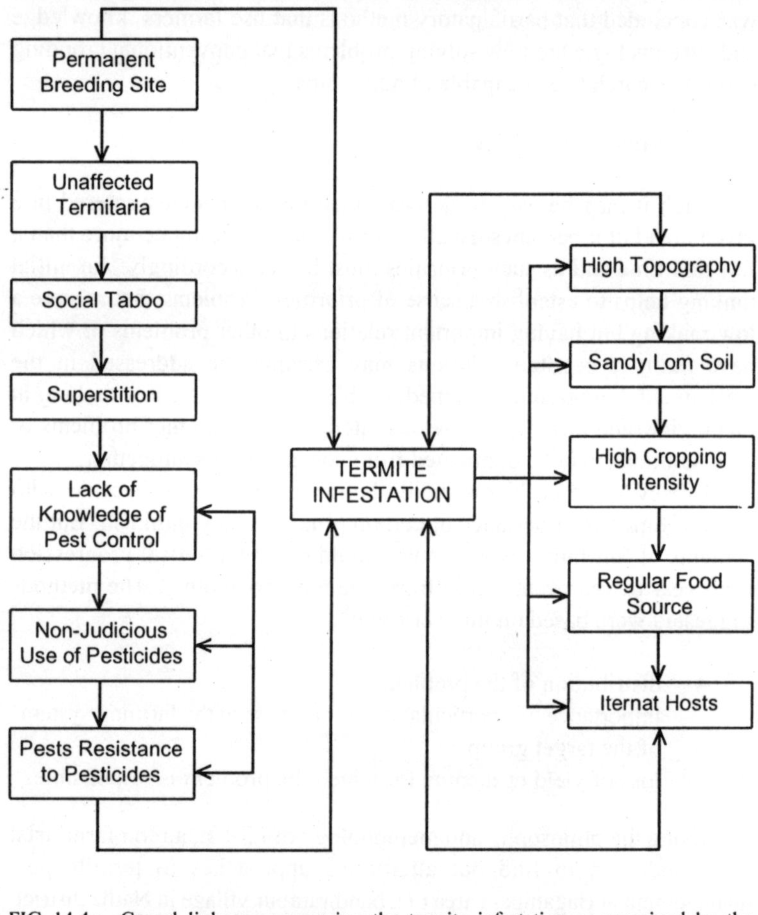

FIG. 14.4 : Causal linkages concerning the termite infestation as perceived by the farmers of Baganipara, West Bengal, India.

— to develop alternative method of termite pest management for sustainable and productive agriculture.

But to achieve these, prime importance should be given to design agricultural systems that are not only less polluting but intrinsically more efficient and productive too, utilizing the natural biological processes and the resources available to us.

On the basis of the results achieved during 'On-Station' trial on the antifeedant and/or repellent property of Elephant-Foot Yam (Amorphophallus campanulatus fam: Araceae, var-'Kavour') was explored in the 'On-farm' condition through farmers active participation with following experimental details.

1. (i) Number of treatments—3
   (ii) Number of replications—10
   (iii) Number of control—1
   (iv) Experimental design followed—RBD (factorial)
2. Casual linkages and system interactions (vide Fig. 14.4)
3. Missing information:
   (i) Indigenous tree hygiene practices
   (ii) Farmers' perception on advantages of termites in terms of soil aeration and nutrient enrichment.
4. Hypothesis formulation: crop replacement as one of the measures to escape termite infestation.
5. Nature of trial: Researchers design and farmers implement.
6. Data recorded on:
   (a) Variables —% damaged plants
   —Cost of manures and fertilizers
   —Cost of seed
   —Cost of irrigation
   —Cost of pesticides
   —Cost of labour
   —Crop yield
   (b) Periodicity: Once a week till harvest of the crop.
   (c) By whom: Field investigators and monitored by the researcher
7. Experimental analyses made on:
   —Agronomic: yield of crop
   —Statistical: RBD (factorial) + CD

—Economic: Comparative cost advantage and risk aversion.
—Farmers' evaluation: Matrix ranking.

Results and Discussion

The results achieved on the antifeedant and/or repellent effect of Kavour variety of 'A. campanulatus' following crop replacement hypothesis were very interesting and encouraging to the farmers of Baganipara. To control termites the farmers have so far used the soil-treating insecticides, mostly of organochlorine group but with discouraging results, while 'A. campanulatus' acted so nicely as antifeedant and/or repellent crop against the same pest that the level of infestation was found to be zero (0) in all treatment combinations. Besides, the comparative cost advantage and risk aversion analysis indicated profitability of growing this crop as antifeedant and/or repellent against termites.

The commercial pesticides have undoubtedly helped reducing crop losses due to pests to a remarkable degree. But these pesticides have also been reported to be responsible for the major environmental and health problems of today e.g., reduction in the abundance and diversity of wild life, human health hazards associated with chronic or acute exposure to dangerous products in the work place and contaminated food, water and air (Conway and Pretty, 1991 and Gips, 1987). Repeated applications of these synthetic pesticides have selected 'pesticides resistant pests' world-wide and at the moment about 450 spp. of weeds have been found to develop resistance against one or more products. Besides, these the death of natural enemies has led to development of newer pest species which were not known earlier. Thus the impression is that more and more pesticides have to be used to achieve less and less.

The results achieved by the present author following FSR/E approach to combat termite problem of Baganipara area of Chandirampur village in Nadia district of West Bengal through replacing termite susceptible crops with those of resistant ones has generated much interest amongst the farmers facing such problem, to grow such crops both from termite pest management and economics of crop cultivation point of view. Further more, this type of approach would also help minimise the evils of pesticide use and as such be considered while designing 'Integrated Pest Management' strategies against termites for sustainable and productive agriculture.

Limitations of FSR/E Approach

While going for implementing FSR/E approach to the sustainability of agriculture, its limitations should be kept in mind as.

- the development of new technologies through FSR will have at best only a very limited effect on sustainability problems that are global in nature e.g., change in climate, depletion of sources of external inputs and changes in global biodiversity.
- the impact of any resource—conserving techniques that may come up through FSR will be limited unless, unlike in the past, these interventions are rapidly and widely adopted. Further developments are necessary in FSR/E methods in order to contribute more fully to sustainability in agriculture.
- a stronger sense of dynamics and trends is needed in all stages of FSR, including those of diagnosis, experimentation and assessment.
- the FSR/E team needs to become more familiar with processes of resources degradation and their underlying causes. Some of these causes may involve externalities, common property resources, property rights and the distribution of assets, public policy and other issues where the skills and specialities of social scientists can be brought to bear.

However, by insisting on realism when confronting sustainability questions, FSR/E approach can still cast itself in starring role in research on sustainable agriculture.

REFERENCES

Chambers, R., 1986. Normal Professionalism, New Paradigms and Development. IDS Discussion Paper 227. Sussex University, England.

Collinson, M.P., 1987. *Farming Systems Research: Procedures for Technology Development,* International Maize and Wheat Improvement Centre (CIMMYT) Mexico City, Mexico.

Conway, G.R., 1985. Agroecosystem Analysis, *Agricultural Administration*, 20, pp. 31-55.

Conway, G.R., and Pretty, J.N., 1991. Unwelcome Harvest *Agriculture and Pollution*, Eastern Publications Ltd., London, p. 645.

Gips, T., 1987. Breaking the Pesticide—Habit. International Alliance for Sustainable Agriculture, Minnesota, p. 372.

Lightfoot, C., De Guia, Jr O., and Ocado, F., 1988. A Participatory Method for Systems—Problem Research: Rehabilitating Marginal Uplands in the Philippines, *Expl. Agric.*, 24: 301-309.

Lightfoot, C., Pingali, P., and Harrington, L., 1993. Beyond Romance and Rhetoric: Sustainable Agriculture and Farming Systems Research. *Eastern India FSR/E News Letter*: 7(1): 18-21.

Spedding, C.R.W., 1979. *An Introduction to Agricultural Systems.* London, Applied Science Publishers.

15

WATER RESOURCES OF THE AJAY RIVER BASIN
A Study towards Sustainable Development

MRIDULA NIYOGI

Introduction

WATER becomes a scarce resource for its multiple use with manifold increase in economic activities. It is more so as its availability is location-specific and conditioned by various factors interacting with each other in a complex manner that leads to a well defined natural system—the hydrological cycle. This cycle is a fairly large natural system of the world that operates through the atmosphere, lithosphere and hydrosphere. The path taken by water through the lithosphere is of vital importance to man. In the process of utilisation of water, man modifies both its quantity and quality at different magnitudes in various phase of the cycle. The extreme sensitiveness of the system to human intervention calls for proper understanding of the processes involved, so that its rational utilisation is possible.

Sustainable development emphasizes on a production system, the input-output ratio and its impact on the bio-physical system (Chattopadhyay & Carpenter 1991). Thus it is location-specific and production-specific. But it is also admitted that one of the major task to approach sustainable development is to gather knowledge regarding the dynamics of ecosystem and the identification of the elements causing interruption to this system. Then comes the question of mitigating the extent of interruption by suggesting development alternatives.

In this chapter an attempt has been made to understand a riparian ecology for water is considered as a "limiting leading input" in a rice based economy (Chattopadhyay 1977). Subsequently the understanding with adequate data-base has been gainfully employed to quantify the water resource situation along with the different components of the basin hydrological cycle. In the perspective of existing pattern of utilization of water, the guidelines for future development have been suggested.

Riparian Ecology of the Ajay

The Ajay is one of the most important right bank tributaries of the Bhagirathi (a distributary of the Ganga) in the lower gangetic flood plain. The river with a length of about 299 kilometres rises from the eastern fringe of the Chotanagpur Plateau of Bihar and after traversing in a east-south-easterly direction it joins the Bhagirathi at Katwa located at about 216 kilometres upstream of Calcutta. The basin has an area of about 6221 kilometres (extending from 23°25′ to 24°35′N latitudes and 86°15′ to 88°15′ E longitudes) of which 57 per cent lies in Bihar and the rest is in West Bengal. The basin stretches in between the Mayurakshi Basin on the north and the Damodar Basin on the south for which the lower part shares the benefit of canal irrigation from both the river valley projects.

In evaluating riparian ecology six principal factors viz. lithology, physiography, climate, soil, natural vegetation and landuse that determine the availability of water both surface and underground in a basin have been taken into consideration.

Geology

The area is underlain by varieties of rocks. According to the water retaining and transmitting capacities of the rocks, the following three litho-units have been identified in the basin.

(i) *Consolidated formation of the Archaean Gneiss:* This high grade metamorphic rocks extensively found in the upper part of the catchment and it covers 61 per cent of the total basin area. Groundwater is confined mainly in joints, fractures and faults creating secondary porosity in the formation and in weathered residuum. There is no permanent water table that contain the river flow in dry period. Rather rainwater is stored in valley-fill deposits.

(ii) *Semi-consolidated Formation of the Gondwana:* This formation crops out in the Sahajori and Jainti coalfield in Santhal Parganas and Andal-Raniganj coalfields of Burdwan. Covering 8 per cent of the basin area, it consists of a massive sedimentary formation of sandstone, grit, conglomerate, shale and ironstone shale with intervening layers of coal seams. Sometimes the thickness reaches upto 2100 m in the Raniganj coalfield.

(iii) *Unconsolidated formation of the Alluvium:* Rest of the basin is covered by the sediments of marine-estuarine-fluviatile origin deposited in a groundwater basin demarcated by a fault scarp in Post-Gondwana Period. The thickness of the alluvium increases towards east and southwest.

Relife

Physiographically the area is characterised by (i) erosional plain studded with monadnocks developed on the gneissic terrain in the upper part of the catchment. Vigorous gullies that sometimes reach to a depth of 6 metres are not uncommon; (ii) erosional plain with broad swells and depressions but outstandingly devoid of monadnocks developed primarily on the Gondwana sedimentaries and partly on the gneissic terrain; and (iii) riverine aggradational plain extensively developed in the formation of fans which merges with the Bhagirathi Delta proper.

Climate

The basin experiences monsoon type of climate and rainfall is concentrated in a period from mid-June to October. During this period, heavy rainfall is associated with depressions and cyclones traversing the lower Gangetic plain. Mean annual rainfall of the basin is 130 cm, 85 per cent of which precipitates during the rainy season. The average annual temperature is around 26°C with average of minimum (19°C) in the month of January and the average of maximum (32°C) in the month of May. Mean annual rate of potential evapotranspiration (PET) is 145 cm with marked seasonal variation. In fact the whole basin enjoy a somewhat similar climatic condition and although rainfall shows more spatial variability than any other climatic variables, yet it does not exceed 30 cm, on an average. Co-efficient of variation of annual rainfall ranges from 15 to 25 per cent indicating the chances of occurrence of flood and drought.

Soil

The spatial orientation of soil closely conforms to the basic structural layout of the plateau and the plain. Red and Yellow and Red soils have been formed on the Archaean gneiss under tropical climate. Due to prolonged pedegenesis the soils are devoid of bases and are slightly acidic. Concentration of hydrogen ion has smaller aggregating effect on colloidal clay. This coupled with low content of clay and organic matter (due to indiscriminate deforestation) have made these soils susceptible to erosion However, in the lower part of the basin, both the texture and fertility status of the soils improve.

Landuse

The extent of man-nature interaction is manifested in the pattern of landuse. The percentage of cultivated land is 60 to 70 even in the blocks located in the upper-most part of the basin. Cultivable waste land measures upto 10 per cent in most of the blocks and forest cover on the other hand dwindles to only 9.9 per cent on an average. Dry peninsular type of Sal (*Shorea robusta*) forest predominates the area.

Analyses of various factors that constitute the riparian ecology of the Ajay reveal that two natural sub-systems, one geotectonic-structural and the other climatic are in operation. Interactions of these two processes give rise to physical landscape, soil and vegetative cover. Human activities initiated in this natural ecological set-up, influenced the landscape in turn and became a part of the process, interrupting the continuity of the natural system in the process of utilisation of land and water.

In this basin, the mode of occurrence of groundwater is primarily determined by the particular lithological formations and geotectonic setting. Similarly, physiography exerts its control on the distribution of surface water. Higher magnitude of slope in the upper catchment of the river is favourable for quick run-off that accumulates in the lower part of the basin and drains out slowly through the trunk stream as the gradient becomes very gentle. Extensive development of flood plain enhances the capacity of storage by promoting percolation. As a result favourable soil moisture balance is available for crop growth.

Atmospheric precipitations is the single source of water and determines the total resource endowment of this basin. Being a part of the atmospheric circulation of the monsoon land, the climate is more or less an independent component of the basin hydrological cycle. The extremes of the variability of rainfall are manifested either as floods or as droughts.

Thus it becomes evident that the spatial and temporal variations in water resource endowment may be explained by the inherent processes operating within the riparian system. The sloping terrain of the upper part of the basin is deprived of both the surface and groundwater and availability of water increases down the basin as the progressively productive lithological formations are encountered along with the increasing valley storage capacity.

Human Intervention

The basin, particularly its lower part, witnessed some of the earliest agricultural settlements of the riparian tract of West Bengal (Biswas 1984). This agricultural community developed an unique system of "overflow irrigation" as early as 3000 years ago suited to meet the special need of the area as basin irrigation suits those of Egypt or "perennial irrigation" meets those of Babylonia (Willcocks 1930). There are ample evidences that the civilization flourished along the course of the Ajay Traded with foreign countries and the river was one of the main navigation arteries of Bengal at that time (Das Gupta 1964).

During the colonial period, the Permanent Settlement was enacted in 1793. The land-laws of this act were not only detrimental to private initiative in maintaining such a large-scale operation, but also gave an incentive towards the reclamation of land for agriculture at the expense of forest cover. As a result, within a span of hundred years, more than 50 per cent of the basin came under plough in Birbhum, Burdwan and Santhal Pargana districts. The situation is well portrayed in the report "what about half a century back was thick jungle and heavy plains of grass is now almost a sterile and barren waste" (Gastrel 1983).

Another important event took place in the mid 19th century i.e., the commercial exploitation of coal in Kasta during 1840-47; in Jainti during 1886-87; in Trans-Ajay area during 1907-1908. The opening of the new collieries initiated the extension of railways that on the other hand had profound effect on forest clearing as it made the interior places accessible. The impact of these events on the riparian ecology was three fold : (i) the mining operation and settlement site, (ii) forest felling went on with an increased tempo and surrounding 'Sal' (*Shorea robusta*) was exhausted soon due to its wide demand for railway sleeper, and (iii) underground working in the mines created interruption in geohydrological system as there was continuous withdrawal of water to carry the mining operation and the abandoned mines acted as vast underground reservoir as a result of which the contribution of groundwater flow to the base flow of the river reduced considerably. Thus the Ajay, which was an important navigation artery of Bengal even upto the mid-17th century turned into an ephemeral stream within a span of two centuries (Government of Bengal, 1866). Today even during the monsoon period the river cannot sustain a stable water regime. Rather, the high discharge is associated with the recurrence of sudden flood freshets.

With the decay of the ancient irrigation system that was also a preventive measure against flood, coupled with pronounced change in landuse pattern, the basin came under the grip of two environmental hazards like soil erosion and flood. Construction of railways and roadways further aggravated the situation when drainage congestion was created particularly at the cross-points. So, embankments were built by local landlord to prevent ingress of flood water in mid 19th century. Till today it is the only preventive measure against flood and the Irrigation Department has reinforced the embankments only.

The magnitude of soil erosion can be comprehended from the sediment load of the river that is 1.01 kilogram/m^3 at Nutanhat (a few kilometre above the confluence with the Bhagirathi). The contribution of the Ajay to the total sediment load of the Bhagirathi was 5.5 to 21.3 per cent during the period from 1962 to 1972. The percentage has increased substantially by about 15 per cent from 1968 onward (Port Commissioners 1974).

Available Water Resource

Surface Water

The river regime is marked by extreme emphemerality in tune with the monsoonal rainfall. It is of AW type which represents "tropical single maxima with a long low water period from four/five to six/seven months" (Bakinsale 1969). Mean annual discharge of the Ajay at Nutanhat is 65.46 cumecs (8-yr mean) that rises to about 247.95 cumecs in August and declines to 13 cumecs in November and ultimately less than one cumec in March/April. Extreme annual variability of riverflow vis-a-vis water resource situation can be comprehended from the Table 15.1.

It may be seen in the Table 15.1 that the yield from the river in drought years is almost half of that in the flood years. Under this situation the simple mean value of river flow does not help a planner. So water yields at different probability levels have been worked out in Table 15.2 with the help of two parameter Gamma probability model (Mooley and Despande 1980).

TABLE 15.1: Average water yield of the Ajay and its tributaries for monsoon period in normal, drought and flood years

(*in Million Cubic Metre (MCM)*)

Site at	*Punasi (Ajay)*	*Siktia (Ajay)*	*Nutanhat (Ajay)*	*Burhai (Pthro)*	*Dhakwa (Jainti)*
Yield from Mean flow	133.36	1230.34	1762.17	147.54	129.16
Drought year (Av. of 1960, 66, 72)	84.73	855.83	1206.63	129.00	102.59
Flood year (Av. of 1970, 71 & 73)	158.99	1980.57	2974.29	153.71	181.15

TABLE 15.2: Yield of surface water in monsoon period (June-October) at different probability levels

(*in Million Cubic Metre (MCM)*)

Stations	*Probability level in per cent*				
	1975	*1980*	*1985*	*1990*	*1995*
Punasi	97.70	91.15	83.91	75.38	64.05
Siktia	1031.66	979.14	931.13	873.03	791.57
Nutanhat	1587.81	1549.19	1504.98	1450.57	1372.75
Burhai	110.68	103.84	96.24	87.25	75.01
Dhakwa	99.51	93.86	87.56	80.04	69.73

Ground Water

The availability of groundwater is conditioned by the geohydrological properties of the lithological formations. Diverse lithology in this terrain causes variations in groundwater yield from one unit to another as mentioned earlier.

The prospect of groundwater in consolidated formation is restricted in such areas where the weathered residuum is thick and the open pore spaces are relatively greater. According to an estimate by Central Ground Water Board the prospect of groundwater in this formation is below 5 m^3/hour.

In the semi-consolidated formation of the coalfields groundwater occurs under water table condition in the weathered mantle that form the near surface aquifer. Moreover, the fractures, fissures in sandstone and shale are open enough to transmit water down to a depth of about

200 m. It give rise to deeper semi-confined aquifers which is moderately productive i.e., 5-50 m^3/hour.

The geohydrological picture of this area has been significantly altered due to mining operation. All the collieries in Santhal Pargana are now abandoned with water logged shafts and partly depillared condition (approximate reserve in the Jainti Khas and central collieries is 2.17 million cubic metres). Active collieries in Andal-Raniganj area have profound effect on the groundwater condition. During the mining operation continuous pumping of the seepage water drags the water table downward. In summer it goes down to 17 m below ground. However it recuperates in post-monsoon period upto a depth of 1-2 metres.

The deep and uninterrupted deposits of the unconsolidated formation (alluvium) are good repositories of groundwater. It occurs under water table condition in near surface aquifer at a depth of 3 to 6 metres (Niyogi 1984). Deeper aquifers in which water occurs in semi-confined to confined conditions have been traced at different level even below 100 metres. These aquifers are interconnected both vertically and horizontally. Depending on the prospect of yield the Central Ground Water Board has divided the area into two zones (a) the Western sector coinciding with the marginal condition of the groundwater basin where yield varies from 50-15 m^3/hr and (b) the eastern sector with high yield above 150 m^3/hr. This prospect of yield is related to the deeper aquifers below 52 m tested by heavy duty tubewells.

Existing Pattern of Water Utilisation

The uses of water in the Ajay Basin are mainly consumptive that includes domestic, industrial and agricultural usage in which water is abstracted from its natural sources and conveyed to some other points before being utilised. In its use, most is often consumed and the rest is recycled.

Agricultural Use

Agriculture being the main sector of economy shares 85 to 90 per cent of total labour force and 60 to 70 per cent of the total area of the basin. In an assessment of quantum of water consumed by this sector, the cropping pattern, area under principal crops and their water requirement have been taken into consideration.

Commensurate with the climatic rhythm of this area two main cropping season prevail. The principal one is Kharif (June/July to Nov/Dec) and another is Rabi (Oct/Nov to March/April). Their intermediary is Rabi-summer.

In lower part of the basin there are three rice season viz. 'Aus', 'Aman' and 'Boro' conditioned by the topographical situation. 'Aus' is grown in well drained upland areas where a second or even a third crop is cultivated in Rabi-summer season under favourable soil moisture condition. 'Aman' is cultivated in lower ground where water stands for a longer period. 'Boro' rice is cultivated mainly with the help of irrigation. It should be mentioned here that only 25 per cent of net sown area of the basin enjoys irrigation facility.

A somewhat different cropping pattern is seen on two terraced field in the undulating terrain of the upper catchment. In kharif, drier variety crops like 'marua' (finger millet) 'bazra' (pearl millet) Kodo godli (inferior type of millets) are cultivated in the upper terraces, Maize and upland variety of rice are cultivated in intermediate terraces and transplanted rice in the lower most sequence. In 'Rabi' season wheat is cultivated in lowland with the help of irrigation while other lands remain fallow.

TABLE 15.3: Hectarage, consumptive use and water requirement of the principal crops in the Ajay Basin

Crops	*Area in ha.*	*Consumptive use in cm*	*Water requirement in cm.*	*Total volume lost through consumptive use in million m^3*	*Total volume required for consumptive use in million m^3*
Aus (Rice)	10.200	66.9	150.0	68.24	153.00
Aman (Rice)	2,21,637	79.3	144.5	1,757.58	3,202.65
Boro (Rice)	5,626	66.1	169.0	37.19	95.08
Wheat	18.636	31.5	32.7	58.17	60.93
Maize	7,693	41.2	64.2	31.69	49.38
				1,952.87	3,561.04

Varieties of crops are grown in this basin. Of these cereals, three varieties of rice, wheat and maize dominate the agricultural scenario. Others are insignificant in respect of areal coverage. So, with a view to estimating the water utilised by the agriculture sector, water requirements of the above five cereals have been taken into consideration (Mitra *et al.*, 1979).

Industrial Use of Water

The basin is backward industrially. At present there are two public sector undertaking, one is Chittaranjan Locomotive Works and another is Hindusthan Cables Ltd. at Rupnarayanpur. Their demand is partially fulfilled by transporting water through pipeline from the Maithan Reservoir of the Damodar Valley Corporation along with local storage (Table 15.4). There are two glass and silicate industries at Madhupur in private sector. Total water consumed by these industries is about 0.08 million litres/day, which is withdrawn from the large diameter wells.

Domestic Use of Water

Total population of the basin has been estimated as 2.14 million in 1991 which is expected to reach 2.42 million by the end of this century. Of these, 88 per cent lives in villages where only 0.77 per cent of the total population is served by centralised water supply. Even not all the urbanites could enjoy tap water facility. Only six out of twelve urban centres have centralised water supply. Moreover per capita water supply is far below the desired level i.e., 90 litres/day except Chittaranjan township and Rupnarayanpur.

TABLE 15.4: Water consumed by the industries in the Ajay Basin

	Units	*Sources*	*Consumption in million litres/ day*
1.	Chittaranjan Locomotive Works	Maithon Reservoir and Local Storage (Surface Water)	3.64
2.	Hindustan Cables Ltd.	Maithon Reservoir (Surface Water) Local Sources (Groundwater)	0.73
3.	Glass & Silicate Industries	Wells (Ground Water)	0.08
		Total	4.45

Total quantum of water consumed by the rural people in this basin has been estimated on the basis of a survey conducted in other parts of India (Nag & Kathpalia 1975). According to that survey 25l/day/capita water is consumed in areas where there is no centralised water supply. On that basis total quantum of water consumed in rural households was

46.5 million litres/day in 1981 and the demand is expected to reach 60.5 million litres/day in 2001.

On urban side, 27.24 million litres of water is catered daily to the six urban centres. People of the other urban centres meet their demand from wells and tubewells. It has been estimated that total urban population of this basin is expected to reach 0.39 million in 2001. So total water requirement will rise to about 35.1 million litres per day to meet the demand.

TABLE 15.5: Utilisation of water for various purposes in the Ajay Basin

Purposes	*Volume in million m^3*
Agricultural	3561.04
Industrial	1.63
Domestic	26.92
Total	3579.59

Uses of water for various purposes comes to about 3579.59 millon cubic metres (MCM) approximately (Table 15.5). This amount constitutes 45 per cent of the total water endowment of the basin (7913.11 million m^3), if the additional water from the adjacent basin which is negligible is not considered. Agriculture being the main sector of economy takes the lion's share (99%) of which 54 per cent is used consumptively and the rest is recycled. Similarly, a significant proportion of Industrial and domestic waste water also take part in recycling process.

The Basin Hydrological Cycle

It has been possible to approximate the quantitative behaviour of different components of the hydrological cycle operating in the basin from various analyses and investigation (Niyogi 1988). This model takes into account total water input, loss by evapo-transpiration, run-off, percolation, the use of water and their inter-relationships (Fig. 15.1). The basic assumption behind this model is as given below (Hore 1969):

$$P = ET + CR + dS$$

where P = Precipitation, ET = Evapo-transpiration, CR = Basin Channel run-off, dS = Change in Storage.

In this diagram total water endowment has been estimated from the average annual rainfall of the basin multiplied by its area. Similarly the

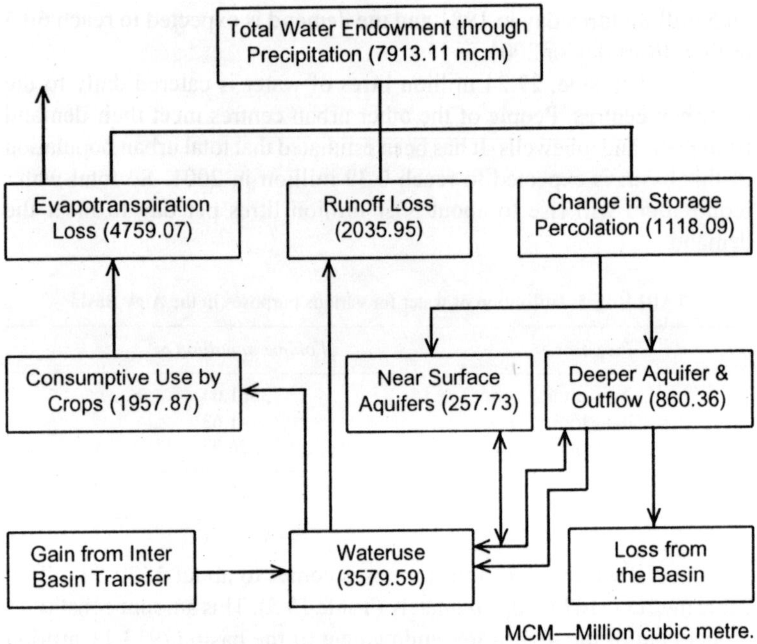

FIG. 15.1. The components of the hydrological cycle in the Ajay Basin

volume lost by evapo-transpiration has been calculated from the average annual rate of evapo-transpiration in this area. The measured discharge at Nutanhat (there being no other discharge measuring site downstream of it) has been taken as indicative of the annual run-off of the basin. This approximation is thought to be valid, since there is no sizeable run-off or base flow contribution below this location. In absence of any direct measurement, the volume of percolation has been obtained by subtracting the volume of run-off and that of evapo-transpiration from the total input.

Percolated water brings a change in ground water storage both of confined and unconfined aquifers. Change in storage in near surface aquifers has been assessed from the fluctuation of water table (Niyogi 1984). A part of percolated water goes to the deeper aquifers or outflows from the basin through leakage. The loss through leakage depends on the continuity of the permeable lithological formation outside the basin. These two components have not been quantified due to inadequate data.

It is thus seen that about 60 per cent of the total water endowment is lost through evapo-transpiration and 26 per cent of the total is carried off the basin as channel run-off. The rest i.e., 14 per cent percolates into the soil. On the utilisation side 3579.59 MCM of the water are used by the agricultural, industrial and domestic sectors. Of the total volume shared by the agricultural sector about 5 per cent is used consumptively by the crops as already mentioned and the rest is recycled among various components of the cycle through a series of transfer. As a result, the resource vista is enlarged. Although the diagram simplifies and approximates the situation, the system in reality is much more complex.

Concluding Remarks

The study of riparian ecology and its degradation on one hand and the existing pattern of water utilisation on the other generate some issues that need consideration while planning towards sustainable development. It is obvious that the present pattern of land vis-a-vis water utilisation is no longer tenable by this ecosystem and that is manifested by the intensification of natural hazards like soil erosion and flood. So an integrated approach for water resource utilisation is necessary and this approach should consider the following aspects.

Restoration of Riparian Ecology

To arrest soil erosion and excessive run-off in the upper part of the basins extensive afforestation programme on the watershed region along with the marginal land should be undertaken. Side by side other anti-erosional measures like check dam, contour bund etc. and eco-friendly agricultural practices like contour ploughing, contour terracing etc. should be introduced. To release excessive pressure on land poultry and animal husbandry may be encouraged as an alternative source of income.

In the flood prone areas, flood plain zoning is the most desirable device in long term planning. However flood jacketing with embankments along the meander belt instead of along the river course may be taken as an immediate measure against flood havoc (Niyogi 1987). It will atleast reduce the chances of breaching the embankments.

The disused collieries that are now acting as vast underground reservoir should be managed scientifically.

Preservation of Existing Water Rights to the Beneficiaries

Under the doctrine of prior appropriation, the first user of water acquires a priority right (Fowler 1960). So water resource planning should regard the water rights of the beneficiaries as far as possible. It is particularly important in the present context as the use of water in the upstream area adversely affect the beneficiaries of the downstream. A specific example could be cited from Santhal Pargana where the subsurface flow through the valley fill deposits of the Ajay and its tributaries are the substantial source of water round the year in the midst of the hard rock terrain. Presently the department of Public Health Engineering (PHED) and Railway are withdrawing their required water (2.5 and 0.068 million litres/day respectively). But PHED of the lower riparian state i.e., West Bengal and the other departments are also planning to tap the sub-surface flow of the Ajay at various reaches. All these schemes after completion will require 33 million litres/day even in the driest month and that amount may not be available in sub-normal rainfall year. Individual plan has been chalked out on the basis of trial boring at the sites at different times. Coordination among all the departments is urgently necessary for the sharing of this limited water resource and before going for any planning regarding the apportionment of water. Priority should be placed to meet the demand of potable water that is expected to reach 36 MCM by the end of this century.

Stepping up of overall economic growth by the intense and optimal use of water: It is seen in the Basin hydrological cycle that we are utilising only 45 per cent of the total water resources. About 26 per cent is lost from the basin as surface run-off and 14 per cent is stored as ground water. These together constitute 40 per cent of the total. There is no surface water scheme expect a medium irrigation project on Hinglow river (a small tributary in Bihar-West Bengal border) and a large portion of ground water is still unutilised. So there is scope to develop further both the agriculutral and the industrial sector for which an integrated approach aiming at conjunctive use of both surface and ground water becomes imperative.

It should be mentioned here that the study is an appraisal of water resource in perspective of riparian ecosystem. It is hoped that this may provide some guidelines for the planner so that the developmental programme would be more eco-friendly. Certainly such projects in hand would require an exhaustive data-base on micro drainage basin.

Acknowledgement

I am indebted to Prof. Kanan Gopal Bagchi Retd. Head of the Department of Geography, Calcutta University for his valuable guidance in this study. The U.G.C. has financed the scheme.

REFERENCES

Biswas, A., 1984. History of Northern Rarh—A Geographical View. *Geographical Mosaic* ed. by S.C. Mukhopadhyay, Modern Book Agency, Calcutta, pp. 765-66.

Chattopadhyay, B., 1977. *Water, Cereal and Economic Growth*, People's Publishing House, New Delhi, pp. 57-59.

Chattopadhyay, S. and Carpenter, R.A., 1991. Sustainable Development: Scientific Jargon or a Practical Management Alternative? *Annals, Nat. Asso. Geog. India*, Vol. XI, No. 2, p. 3.

Das Gupta, P.C., 1964. The Excavation of Pandu Rajar Dhibi, *Bulletin of the Directorate of Archaeology*, West Bengal, No. 2, pp. 1-30.

Fowler, F.J., 1960. Some problem of water distribution between East and West Punjab, *Geog. Rev. Amer. Geo. Soc.*, New York, Vol. 40, No. 4, p. 583.

Gastrel, J.E., 1863. Report on the Bankura Forest. Record of the Bengal Government.

Govenment of Bengal (1866). Statement of Navigable Rivers, Canals and Khals in Bengal, Calcutta, p. 38.

Mltra, B.N., *et al.*, 1978. Water Management of Rice Under Different Land Situation of West Bengal. *Rice in West Bengal*, Vol. 1, Directorate of Agriculture, Govt of West Bengal, pp. 71-85.

Mooley, D.A. and Despande, V.R., 1980. Rainfall Probabilities in Karnataka during Summer Monsoon, *Geog. Rev. Ind.*, Vol. 42, No. 2, pp. 120-24.

More, R.J., 1969. The Basin Hydrological Cycle. In *Water, Earth and Man*, ed. by R.J. Chorley, Mathuen & Co. London, pp. 67-71.

Nag, B.S. and Kathpalia, G.N., 1975. Water Resource of India. *Proceedings 2nd World Congress*, Water Resource New Delhi, Vol. II, pp. 379-82.

Niyogi, M., 1984. Ground Water Resources of the Ajay Basin. In *Geographical Mosaic*, ed. by S.C. Mukhopadhyay. Modern Book Agency, Calcutta, pp. 166-83.

Niyogi, M., 1987. Flood Frequency and Magnitude Analyses in the Ajay Basin, *Ind. Jour. Landscape Systems & Ecological Studies*, Vol. 10, No. 1, Calcutta, pp. 28 29.

Niyogi, M., 1988. Water Resources of the Ajay Basin—A Geographical Hydrological Study. Unpublished Ph.D. Thesis University of Calcutta.

Port Commissioners, 1974. The River Ajay—A Study of Hydrological Character, Unpublished report, Port Commissioners, Calcutta.

Willcocks, W., 1930. Lectures on the Ancient System of Irrigation in Bengal. Univ. of Calcutta, pp. 1-5.

16

GEOMORPHIC ANALYSIS FOR SUSTAINABLE LANDUSE DEVELOPMENT
Case Study of A South Kerala Drainage Basin

MAHAMAYA CHATTOPADHYAY

Introduction

SUSTAINABLE landuse development has been a central theme of development initiatives in recent years. In the wide ranging debate on planning for Sustainable Development landuse occupies an important position, after all, sustainability can be primarily conceived in terms of renewable natural resources like land, water and biomass. Given this premise at the outset sustainable landuse development can be acheived through prudent management of land, water and biomass resources.

By definition, sustainable landuse is a type of land management practices that is ecologically conformable, economically viable and socially acceptable. By ecology it is meant the landscape ecology, which represents an integrated unit evolved out of interaction of land, water and biomass. While economic and social aspects are anthropocentric and hence amenable to management practices, the landscape ecological parameters are governed by certain laws of nature. Therefore, it is imperative to learn these natural laws for devising eco-conformable landuse pattern. In case of landscape ecology landform being the interface holds the key to understand the ecosystem. So it is pertinant to investigate the evolutionary processes contributing in shaping the landform. Which in other words is the first step in learning the natural laws active in the area under study. As these natural laws operate within the framework of natural systems, it is perhaps appropriate to investigate in terms of natural units.

Drainage basin forms a natural unit of landform, where the operative denudational and aggradational processes are explicitly manifested. It functions as a limit of erosion, that gives expression to relief, and as a local erosion and denudation base limit (LEDBL). Further, dissection of denudation centres generally lead to planation surfaces. From the source to mouth a river basin exhibits various types of landforms in response to geological structure, processes operative therein and the stage of land building activities. At the micro level, the geomorphic features vary from the ridge top (denudational centre) to the valley bottom (local base level). A cursory glance to any river basin will bring out the different landuse pattern in consonance to the land and water distribution matrix. This is evident in the micro level also.

Kerala, with its high density of river basins and physiographic characteristics including high gradients, manifests a number of geomorphic features different from the rest of south India, and hence offers an excellent opportunity for the study of landform evolution and interrelationship between geomorphology and landuse with the drainage basin as focal point. Once this relationship is understood it is possible to design a sustainable landuse pattern. Keeping this in view, the author has carried out investigation in a relatively small river basin, namely Ittikara, in southern Kerala, and the findings are presented in this chapter.

Study Area and Methodology

The Ittikara river, flowing through Quilon district originates from the Yerur reserved forest area of Kallyani Kunnu (436 metres above mean sea level) of the Western Ghat hill ranges. Its major watershed bifurcates the river Kallada and Vamanapuram from the Ittikara river. This drainage basin is comparatively small with a catchment area of approximately 750 km^2. It is a 6th order basin with 2 sub basins of 5th order, 18 fourth order basins, 60 third order and 344 second order and 1024 first order sub-basins respectively. The landform of the basin broadly presents an undulated ridge and valley topography. Elevation range is not high, but the dissection index of the terrain is high.

In order to analyse the geomorphic development, a 4th order basin of Kulanji Todu, located in the north-western corner of the catchment area has been selected for the detailed study. It has an enclosed area of 48 km^2 with 3 third order sub-basins, 15 second order basins and 73 first order basins. The typical representation of midland topography is distinctly observed. The stream (Todu) originates from the upland of Munnar at an elevation of 169 metres.

Interpretation of landsat imagery, aerial photo, toposheets of 1 : 50,000 and 1 : 250,000 scales and field verification provided the basis for this study. Morphometric and morphographic parameters for the entire basin area have been computed to facilitate the work.

Lithology and Drainage Pattern of the Ittikara Basin

The rock types exposed in this area include Precambrian, crystallines, Neogene and recent to sub-recent sediments. The oldest and the most widespread rocks are the Precambrians, consisting of migmatised khondalites, charnockites, patches of garnet-biotite gneiss and garnet-quartz-feldspar neosomes. Cenosoic sediments occur all along the coastal plain. Blanket-type laterite occurrence is seen in the north-western sector of the basin as an elongated patch forming low flat-topped hills and ridges. Isolated laterite cappings developed over various lithologies are also seen upto elevation of 120 m. The drainage patterns have a complex arrangement in the area, with considerable variation in spatial alignments related to relief, slope and rock type. Dendritic pattern with variable density can be distinguished on a regional scale. In the biotite-gneiss rocks, the pattern is somewhat parallel and angular because of the adjustment of the drainage to subjacent structure. The southeastern tributary Mann Ar flows along a lineament which is indicative from the straight channel. The lower drainage segments form a dendritic pattern with low drainage density along near-flat topography. Kulanji Todu catchment area also displays dendritic pattern, and migmatised khondalite is the major lithology. Small watersheds can be discerned in the interfluve areas. Tributary development is more in the eastern portion than the western counterpart, where dissection index of the terrain is high.

Theoretic Denudation, Erosion Base Limits and Denudation Centres

The denudation process or the concept of a cycle of erosion expresses the evolution of slopes towards a level surface. As a rule slopes of varying inclinations make up the relief, and these are generally convex in their upper part and concave in their lower part. Further, they are limited on one side by the water divide and on the other side by the river bed into which tributary slopes drain. So, the river functions not only as a limit, but also acts as a local erosion and denudation base limit (LEDBL) which is represented as V1 in the landform analysis. The narrow area enclosed between the LEDBL and the summit or denudational centres has been represented as T1 surface in the landform analysis of the basin area (Fig. 16.1). In the Ittikara river basin, it has been marked as the denudational slope area. In this particular area, the denudational slopes around first, second and third order summit areas are narrower than the fourth order summit areas. This also points towards the levelling of terrain with the advancement of stage and processes. The enclosed area by the surrounding streams, whose slopes follow an evolution independent of the neighbouring slopes constitute the denudation centres (Fig. 16.1). In this study we have also noticed that with successive development of hill ridges and flat-concave slope, the denudation centres have become dissected and reduced in sizes. This has helped in the levelling process of the terrain, or in other words regional planation surfaces have started developing. Towards the southern sector, the denudation centres have given way to almost level surfaces through erosion, dissection and deposition.

Potential of the Land Units and Landuse

The 4th order LEDBL is highly enriched with sediment and water followed by the 3rd order and 2nd order LEDBL. The 2nd order LEDBL being indented within the denudational centres are potential sites for water harvesting and moisture conservation. The denudational centres composed of laterites and laterite derived materials act as source zones of sediment and water. The land units and potential landuse types are given in the Table 16.1.

TABLE 16.1: Land types and landuse in a sample drainage basin

Land Units	*Land Potential*	*Potential Landuse*
Denudational centres	Source zone, shallow soil, low water regime, erosion affected	Permanent trees, grass cover settlement etc.
Denudational slopes	Transitional zone, shallow to deep soil, moderate water regime, erosion affected	Settlement and trees, seasonal crops with adequate protection
Planation surface	Source cum sink zone, moderately deep soil, high water regime, low erosion	Settlement, tree crops, seasonal crops in patches
2nd order and 3rd order LEDBL	Sink zone, colluvic alluvium material, moderate to high water regime	Seasonal crops, partly reclaimed
4th order LEDBL	Sink zone, alluvium high water regime	Seasonal crops, partly reclaimed

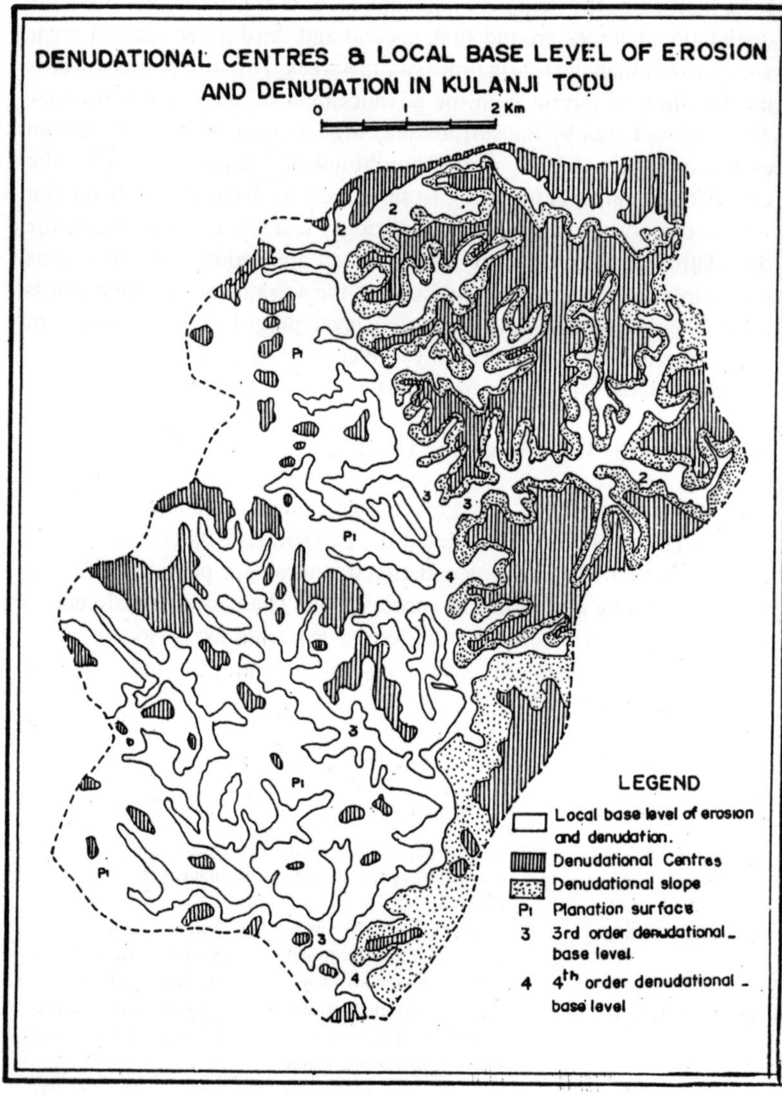

FIG. 16.1: Denudational centres and local base level of erosion and denudation in Kulanji Todu

Conclusion

The results indicates that the shaping of denudation centres is a continuous process which helps to reduce the relief until flat surface is attained independent of spatial and temporal location. The landform map prepared based on slope parameter also establishes this concept. The classification of drainage, interfluves and their ordering are also geomorphologically important, because the dissection and shaping of terrain depends on the disintegration of interfluvial branches situated around summits.

This study helps in understanding the natural laws operative in developing landscape system. Landform landuse interrelationship evolved in this area also provides clues to develope sustainable landuse pattern. Based on the results obtained from this analysis studies can be undertaken to cover similar areas in south Kerala.

Acknowledgement

The author is grateful to the Director, Centre for Earth Science Studies for providing facilities. Grateful thanks are due to Dr.K.Soman, head, RAD for providng the geology map and other facilities. Mr. K. Mohandas has drafted the maps.

REFERENCES

Chattopadhyay, S., Chattopadhyay, M. and Shaji Mohan, 1993. *Terrain analysis of Kerala through remote sensing application.* Part I, Concept, Methodology and Results. Centre for Earth Science Studies, Trivandrum, p. 60.

King, C.A.M., 1976. (Ed.) Landform and Geomorphology. Concepts and History. Benchmark Papers in Geology, 28, p. 403.

Sinha Roy, S. 1979. Laterite Profiles in Relation to Geomorphology in Parts of Trivandrum District, Kerala. Prof. Paper No. 3, Centre for Earth Science Studies, p. 40.

Soman, K. 1980. Geology of Kerala. Prof.Paper No. 8, Centre for Earth Science Studies, p. 62.

Van, Zuidam, 1990. Geomorphologic Development of the Aljuce'n River Basin. *ITC Journal 1990-4*, The Netherlands, pp. 324-34.

17

SUSTAINABLE USE OF LAND
A Strategy for Wasteland Development in Nandivali Village, Pune District, Maharashtra

SUNITA PURKAYASTHA

Introduction

LAND is the most vital resource for livelihood for the majority of people in our country residing in the rural area. They need it for cultivation of food crops, as pasture to feed the cattle, or as forest to provide them with fuel and timber. These needs however conflict with each other because of limited supply of land. Hence, there has been excessive pressure on the land, resulting in improper use, extension of agriculture in marginal land, overgrazing and indiscriminate deforestation, which together has culminated to land degradation. This degrading process is more perceptible in the fragile ecology of hill areas, where the resource on which people depend has deteriorated at an accelerating rate. Forests have been burnt or felled, the vegetation cover removed and steep slopes have been severally eroded. Such degraded land, which can be brought under vegetation cover with reasonable effect and which is currently lying under utilised, and land which is deteriorating for lack of appropriate water and soil management or an account of natural causes are defined as wasteland (NWDB, 1985). This study analyses the problem of wasteland in the Mulshi watershed area of Pune district, Maharashtra. The area falls in the high rainfall zone of Western Ghat. Traditionally the farmers have attempted to increase production through extension of area under crops. Increased pressure on the land has resulted into deforestation and intensification of agricultural activities on unsuitable marginal lands opening up of steep slopes for cultivation without practising conservation measures, increased run off and soil erosion enormously. This has led to land degradation and loss of production base. There has been negligible effort on the part of Government to undertake any eco-development programmes in the area. The problem of land tenure and feeling of insecurity among farmer has further aggravated the situation. Agriculture is mainly subsistence and as such the income derived from it is very meager. Poverty in the area is not linked with landlessness, rather it is related to water scarcity, land degradation, tenure problem and lack of infrastructure.

Objectives

Following are the main objectives of the study:

1. To analyse the landuse of the study area;
2. To determine the causes of wasteland formation in the area;
3. To develop a land capability model for the study area; and
4. To suggest an integrated wasteland development plan for the area.

Methodology

As a case study, the village Nandivali, was selected from the watershed to carry out detailed survey. As the focus of the study is wasteland, the same criteria i.e. area under wasteland from the Pune District census data of 1971 and 1981, was used for selecting the village, Nandivali, which has maximum area under the mentioned category for both the year.

The village level data were collected by participatory analysis. Informal interview both at the individual level with the village elders and also groups of women, children and with the farmers in their field were used to extract key information on the resource base. The secondary data were collected from the land records of Talati and other local government departments like Forest Department and Soil and Water conservation department. The second stage of the study included the classification of village into land capability zones, for which soil samples were collected from selected sites in the village, taking the landuse and slope of the area into consideration. The samples were then laboratory tested to verify the pH, NPK content and texture of soil.

Geographical Background

Nandivali is one of the four revenue village in Mulshi watershed, with one wadi or satellite settlement, Wadwathar under it. The total geographical area of the village is 601.6 hectares and the population is 480 as per 1991 census data, which shows an increase of 86 per cent from 1981 census, figure when the population was only 258.

The general topography of the village is hilly with undulating slopes varying between 2° to 49°. The longitudinal cross-section of the village from the ridge line in the north to the reservoir in the south, gives the following horizontal division, upper ridge or highland, the flat topped plateau or midland, the medium sloping land and the lowland in the bottom.

The village falls in the monsoon dominated area of the Western Ghat. Rainfall is received during the south-west monsoon and 80 per cent of it is concentrated in the first three months from June to August. Average rainfall recorded by the IMD, Pune in the year 1990-91 was 2100 mm. The average maximum and minimum temperature ranges around 32° and 18° celsius respectively.

The soil is lateritic in origin, except, in some flat topped hills where the exposure of basalt is also observed. The colour of soil varies from brownish red to bright red due to preponderance of hydrated iron-oxide. It is acidic in nature with pH ranging from 4.5 to 6.0. The soil is porous and does not have good retention capacity of moisture. The depth of soil varies from 30 cm along the banks of the lowland to 6 cm on the higher slopes and flat topped plateau.

There is no permanent source of water in the village. Only a few small streams are there, which are seasonal having water for only 3-4 months during monsoon. As such they do not fulfil the irrigation needs of the village. Due to steep slopes and excessive forest denudation the moisture absorption of the soil is low and most of the run-off flows down into Mulshi reservoir, owned by Tata Hydro Electric Company. Water table is very deep and can be tapped only at 60 m depth (Benninger, 1991). The two open wells situated in the village run dry during summer.

Agriculture

Nandivali is a typical agriculture dominant village with little support from non-agricultural operations. The main crop grown is paddy which is cultivated mostly in the plain land situated near downstream. The variety of Paddy grown depends on the soil and moisture availability of land. The other major crops are the coarse millets like *Nachani, Varai* and *Til*, which are cultivated on the hill slopes with shallow soil called '*varkas*'. As these lands are of low fertility, it is rather difficult to put them under cultivation every year. Therefore, a pattern of shifting

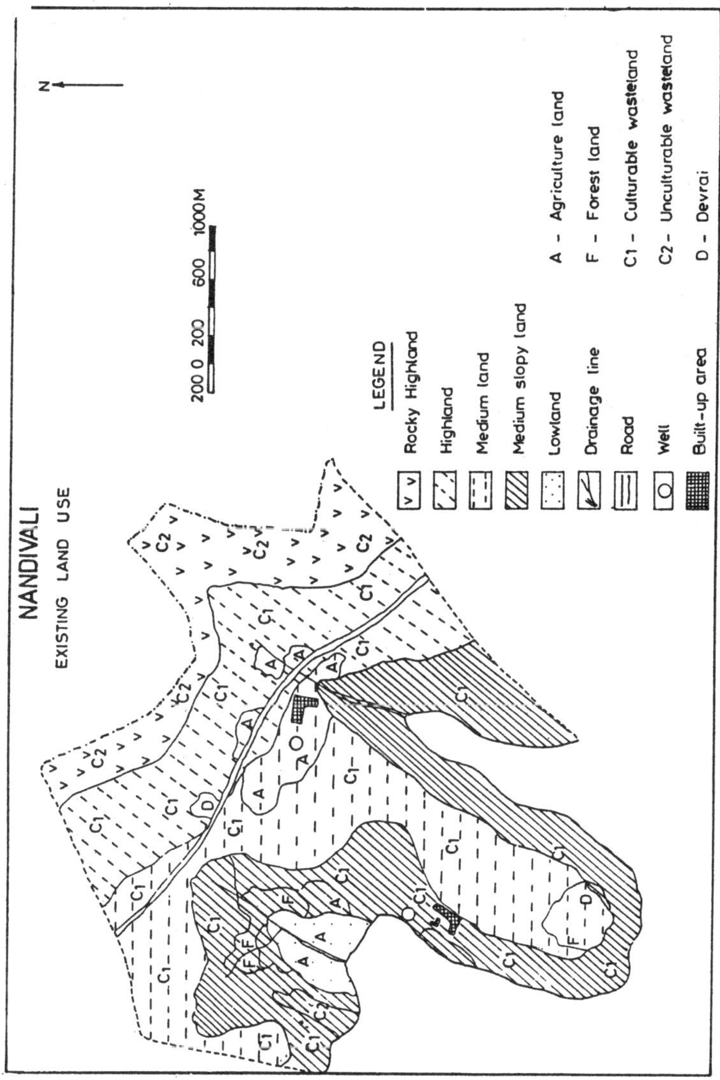

FIG. 17.1: Nandivali—existing land use

cultivation is practised, where the part of land cultivated once is left fallow for 2-3 years. Agriculture is total rainfed due to lack of irrigation facilities. As such there is no scope of growing more than one crop a year.

Land Tenure

The land tenure has a vital bearing on agriculture of any area. It governs the condition under which land is husbanded and inputs are used. Almost 80 per cent of land holding in Nandivali are below 5 ha.

The three main categories of land ownership existing in the village are:

(i) Land in the name of cultivator,
(ii) Land belonging to absentee landlord, and
(iii) Land owned by Tata Hydro-Electric Company.

Majority of the land belongs to absentee landlords, which have been cultivated by tenants for almost 75 years. But even now there names have not been entered into the land records. In the absence of alternative employment opportunities, farmers tend to sow the land with annual crops and often they are not remunerative. The perception of insecurity of land tenure deters the farmer from undertaking any long term investment. Like tree plantation which involves long gestation period. Also due to absence of land rights, they are unable to participate in any land based government programmes.

Land Use Analysis

The analysis of landuse is one of the most important aspects for studying the relationship between man and his environment. It is essential not only to adjust the present landuse to the capability of area, but also to direct the expansion of intensive land utilization into potential area and check the undesirable trends of irrational use of land.

Landuse in the Nadivali village, can be broadly categorised into six types (Table 17.1). Brief description of each type is given bleow:

Forest

Forest area accounts 1.18 per cent of the geographical area only. But field observation revealed that there is hardly any vegetation in the land governed by the Forest Department. It is highly degraded with few shrubs. In addition to this even in private lands there are few patches of dense vegetation and open forest cover.

Devrai

It is a sacred grove dedicated to mountain god and hence prohibited for any form of exploitation.

TABLE 17.1: Landuse types

Landuse category	*Area (ha.)*	*Percentage*
Forest	7.11	1.18
Devrai	1.34	0.23
Net Sown Area	56.40	9.37
Uncultivated Land	142.23	23.64
Culturable Wasteland	191.00	64.98
Built-up Area	3.69	0.60
Total	601.69	100.00

Source: Talati land records.

Culturable Wasteland

This category includes all land available for cultivation or taken up for cultivation once, but not cultivated during the current year or last two years in succession. Therefore, it includes the shifting cultivation areas, degraded pasture and grazing land with or without scrub, constituting almost 65 per cent of the total geographical area of Nandivali. In shifting or fallow system of cultivation, the leaves and branches of trees are lapped, small bushes are uprooted and cover of dry fuel is spread over and are burnt in pieces of land in hill slopes. This process is called 'Rub' and it's mainly done for fumigating the land. Cultivation is practical for 2-3 years, after which the land is left follow to regenerate itself. Previously the fallow period used to vary from 8-10 years, but gradually with increasing pressure of human and livestock population, the follow period has steadily declined to 2-5 years depending upon the land holding. This has not only reduced the vegetation cover but also accelerated the process of soil erosion; as large quantity of top soil is washed away, degrading the land. Moreover the continuous burning of land, slackens the process of regeneration of grasses, thereby depleting the supply of fodder.

Net Sown Area

This category of land includes the present area of land under cultivation, which is only 9.3 per cent of the total area. Out of 56.4 ha of land under cultivation, 25 ha is under rice and 30 ha under coarse grain like Nachni and Varai. Only 1.4 ha of land is under fruit trees like Mango, Jackfruit etc. which is very less compared to the potential.

Uncultivatable land

This category of land, located mostly on the higher slopes are characterized by barren exposed rocks and the sheet rock areas with shallow soil which render them unsuitable for production of any biomass.

Built-up Area

The Built-up area is the residential area of the village.

From the preceding analysis of landuse, it was possible to deduce the causes of wasteland formation in Nandivali which is listed in Table 17.2.

Land Capability Classification

In planning the use and management of land, the first and foremost question to be considered is; what are its possibilities for use? One of the best method for assessing the degree of limitation of present landuse, and also the productivity and potential of land for its proper management in future, is to divide it into categories based on the capability of soil.

Such a system known as land capability classification, has been developed by United States Department of agriculture (USDA) and also FAO, which are universally followed.

According to USDA, the arable soils are grouped according to their potentialities and their limitation for sustained production of common cultivated crops, without specialized site treatment. Non-arable soils (soils unsuitable for long-term sustained production of cultivated crops) are grouped according to their limitation for producing permanent vegetation and according to the risk of damage if mismanaged (Anon, 1980).

Eight classes of land were designated: capability of land decreases from class I to class VIII depending upon the severity of the problem: class I to III are considered suitable from cropland, class IV as marginal and class V to VIII unsuitable for cropland.

The grouping of soil into different classes, sub-classes and units is done primarily on the basis of field investigation of soil properties, slope, degree of soil erosion and climatic conditions.

TABLE 17.2: Wasteland formation, causes and indicators

Problem	*Causes*	*Quantitative Indicator*	*Qualitative Indicator*
Increase in area under wasteland	1. Rate of deforesttation far exceeds the rate of forest generation	Only 1.12 per cent of land is under forest according to the official report.	The Forest Department has not undertaken any afforestation programme in the village. Absence of forest guard in the village. Process of lopping practiced every year reduced the stocking of forest & density of three cover.
	2. Farmers do not have secure legal tenure (wahiwat) of the land they cultivate.	Only five farmers in whole village have wahiwat in their names.	The insecurity of land tenure inhibits the farmers from planting trees & making long term investment in their land.
	3. The fallow period in shifting cultivation has steadily decreased over the years leading to degradation of biomass.	Previously the fallow period was from 8-10 yrs. At present it is 3-4 yrs for medium and large farmers with more than 5 ha of land & 1-2 yrs for small farmers.	The fallow system of cultivation is still economical for the farmer, due to less availability of good agricultural land.
	4. The input required for practising "Rab" has also decreased over the years.	Previously 4 inches of soil used to get burnt in rab, whereas now-a-days only 2 inches of soil is burnt.	The amount of grass used in rab has decreased. The root stock of the grass perishes with periodic burning slowing their processes of regeneration.
	5. The availability of fodder in the village is not enough to meet the total requirement.	Last year 2 cattles died in the village because of starvation.	Amount of fodder bought from the market is also increasing.
	6. Fuel wood availability in the village is on the decline.	Women in the village now have to walk for 4-5 km & spend about 6 hrs a day in search of firewood, especially before the onset of monsoon.	The poor farmers' traditional right of free access to fuel from the land of better off farmers are slowly eroding.

The classification method consists of the following steps:

(a) Identification of indicators and sub indicators;
(b) Evolving a system of weightage; and
(c) Grouping into capability zones.

Characteristics of Land Capability Classes

Class I: These lands have gentle slopes with moderate soil depth and moderately fine texture. They are not affected by any appreciable erosion and soil is workable with medium fertility. Presently this category of land is being used for cultivation of rice and also shifting cultivation. Rest of the area is under dense tree and shrub cover. This class of land is suitable for permanent cultivation with moderate restrictions, as well as horticulture on slightly higher slopes. Contour tillage, inter-cropping of legumes and grains, terracing and levelling, use of green manure and fertilizers are the soil conservation practises which should be useful.

Class II: These category of lands are covered with dense to moderate tree and shrub cover. It is used for grazing and shifting cultivation. They are characterized by gentle to moderate steep slopes with medium deep soil having low fertility and high susceptibility to erosion. These restrictions along with high risk of erosion poses severe limitation to permanent cultivation. Therefore, they can be best utilised by horti-pastoral system of landuse, i.e., combination of high value horticultural crops with good fodder yielding variety grasses. Conservation technique like contour bunding will intercept run-off there by improve moisture availability.

Class III: These lands are covered mostly with shrubs and grasses. Soil cover is shallow, acidic with low inherent fertility. The slope of land varies from 2° in the flat topped plateau region to almost 30′ in the steep slopes of the ridge. This type of land can be properly utilized by adopting sylvi-pastoral system i.e., integrating tree with pasture. Plantation of fuel and fodder trees along with fodder grasses can bind the soil and help infiltration.

Class IV: This category of land has very steep slope. Soil is shallow, highly acidic and subjected to severe erosion.

As such soil conservation measures need to be practised intensely in this area to check further degradation. This type of land can be properly utilised by planting agage, khus-khus and other local variety of grasses, along the contour line in order to stop loss of top soil cover.

Class V: These lands are characterised by very steep slopes with very shallow soil and weathered and exposed barren rock. These limitation poses obstacle for proper and intensive utilization of this category of land.

Proposed Landuse Pattern

The land classification into capability zones indicates possibility of bringing more area under productive uses, through proper management of follow lands, culturable waste and barren land.

The landuse pattern has been suggested on the basis of physical *i.e.*, soil and slope and agro-climatic condition prevalent in this area. Scope for extending the area under permanent cultivation being limited, emphasis should be given on intensive and multiple use of land, through adopting agro-forestry for maintaining ecological stability. Following are the approaches suggested for sustainable use of land in Nandivali.

Agriculture

In order to accelerate the development of agriculture, a co-ordinated approach is required. Beside the supply of inputs like fertilizers and good variety of seeds, emphasis should be given on the provision of agricultural inputs, supporting services and agricultural extension programmes in the village.

More land can be brought under permanent cultivation by terracing, levelling, building and provision of safe diversion channels on agricultural fields along the stream. In the places of shifting cultivation, instead of leaving the land follow to regenerate itself, grasses should be cultivated to revive the fertility of land.

Forest The strategy for development of forest should aim at preservation of existing forest and afforestation and extension of forest in marginal agricultural land, culturable waste and barren land through social and farm forestry. Land with steep slopes between 20-30° and highly susceptible to erosion, quick growing varieties could be introduced. On moderately slopping land indigenous varieties producing minor forest product and fuelwood like Anjan, Sisoo, Shikakai etc. can be grown.

For afforestation of the Forest Department land, located near the settlement, the local communities must be involved by making them equal partners in protecting and managing the state owned forest and selecting species of trees. The local community who will take major responsibility of protecting the trees, and the final produce will be shared by both the partners.

The private lands located on higher slopes with low accessibility could be acquired by the Forest Department on a lease basis for 20-25 years for their effective control, development and management.

Fodder

Fodder development should be taken on an extensive scale for development of culturable wasteland in the village lands having shallow soil cover unsuitable for growing trees. Improved varieties of grasses like style hamata, anjan and marvel can be grown. Ecologically, it will be suitable if areas having slopes of 30° and above are utilised for growing fodder trees, and grasses. In areas having slopes between 10° and 20°, it can be intercropped with horticulture. This can be converted into feed and be stored in fodder banks for consumption during the dry summer months.

Horticulture

The physical and agro-climatic condition of the village is best suited for growing fruit trees with different varieties grasses in areas having slopes of·10°-20°. This will create natural terrace, thereby halting erosion of top soil and retaining the fertility of land. Table 17.5 gives the proposed integrated landuse plan for Nandivali. It enumerates rationally the schemes (input) for the proposed landuse, the class of land where it should be taken up and the actors who will be involved in implementation.

Conclusion

The focus of this study has been on evolving a methodology of land-use planning for proper management of wasteland, in only one village of the watershed. Detailed analysis of the factors controlling landuse is required for undertaking any programme of ecological regeneration.

The socio-economic dynamics of the area, the land tenure problems are government policies on forest land management etc. are some of the important factors which should be considered.

However, any such effort to enrich the village ecosystem will have to began with the people who knows most about their own livelihood system. The local community must be involved in all stages of planning, implementation and monitoring.

The ecological regeneration process should be based on a holistic approach. It should attempt to increase the productivity of all components of the village ecosystem—from its croplands to forest and grazing lands, water systems, animals and above all the people in a way that this regeneration process is sustainable.

REFERENCES

Anonymous, 1980. *National Agricultural Lands Study. Soil degradation: Effect on Agricultural Productivity*, Washington D.C.

Benninger, A., 1991. *Village Ecodevelopment*, Part II, A Methodology for Operationalising the Action Plan, IRD, CDSA.

National Wasteland Development Board, 1985, Wasteland Development Programme, Guideline for action.

18

ECOLOGIC-ECONOMIC EVALUATION OF RENEWABLE RESOURCES FOR SUSTAINABLE DEVELOPMENT
A Micro Level Study

C.K. SASIDHARAN

Introduction

ECONOMIC development, the ultimate goal of which is to improve human welfare, is crucially dependent on the environment and natural resources to provide goods and services which directly and indirectly generate socio-economic benefits. Governments in developing countries are becoming increasingly aware that environmental and natural resource degradation endanger the potential for long term development. The natural resources that are known for certain today is only a part that are actually available. Those which are available are those which can be extracted from nature under the prevailing technological, economic and social conditions (Chapman, 1969). There are different view points as to the optimum human population this planet can support—ranging from 15 billion (Weinberg and Hammond, 1972) to 100 billion (Campbell, 1970). However crude the estimates may be, one casualty of these estimates has been that the quality of environment in which population would be surviving has not been taken into consideration. One thing is certain that as things stand, there are no infinite reserve of resources that can sustain ever increasing human population. The message is simple—conserve and preserve the renewable resources; recycle, improve the technology and process of nonrenewable resources extraction and utilization; reduce the per capita consumption etc.

Mankind's relationship with the environment has gone through several stages. The environment, including natural habitats and ecosystems, provides three main types of services to mankind.

(a) The environment is a source of essential raw materials that support human activities. In particular, natural habitats provide the basis for food and cash crops, fish stocks, forests, domesticated and wild animals, and other natural assets, all of which benefit human society.

(b) The environment serves as a sink that absorbs and recycles the waste products of economic activity, normally at little or no cost to society.

(c) The environment provides irreplaceable life support functions without which life on earth would be drastically altered or even destroyed.

Recently, mankind has begun to look for practical options at sustainable development, that will permit continued improvements in the quality of life with a lower intensity of resource use. Many current problems have arisen either from over exploitation of natural resources or due to poor planning and design of developmental projects. It is necessary to quantify, as far as possible, all the important biophysical and socio-economic changes, likely to result from the project. Hence the key to sustainable resource use is through monitoring systems and information collection, so that the ecologic-economic interrelationships are understood in proper perspective.

Conventional and Environmental Economics: A View Point

Ecological awareness is not new in social history. Ecology is the science which deals with the relationships between all living organisms and their total environment. Protection of biological diversity is a key aspect of the ecological view. The economic approach is to maximize flow of income that could be generated while maintaining or increasing the stock of assets or capital. Mankind began interfering with nature immediately after his evolution. Man's approach to environment and his concept of resources have a temporal element linked to the technological advancement and a spatial element depending upon resource availability. Human welfare is crucially dependent on the environment and natural resources to provide the goods and services which directly or indirectly generate socio-economic benefit. Environmental economics helps us to incorporate ecological concerns into the conventional frame work of human society.

According to Repetto *et al.* (1989), the Classical Economists treated income as a return to three kinds of assets: natural resources, human resources and man-made capital (land, labour, capital). The Classical Economists perceived natural resources and economic development as inter related to each other. For instance, Malthus predicted a stagnation through "absolute scarcity" of land resource which is characterised by diminishing return in the agricultural output after the absolute limit of the land availability is reached. Ricardo was concerned with "relative scarcity" in which both the cost of production and output will keep on increasing and argued that the diminishing returns start from the beginning (Bernett and Morse, 1963). Though Ricardo predicted a temporary scarcity phenomenon, he was of the view that land is heterogeneous in nature (fertility differs among different parcels of land with different quality and agricultural operation). Mill not only accepted the Ricardo's "relative scarcity" of land resource but also brought the non-renewable resource into his analysis (Barnett and Morse, 1963; Barbier, 1990). So in one way or other Classical Economists were concerned with the resource scarcity and its impact on economic growth.

Environmental economics play a key role in identifying efficient natural resource management options for sustainable development (Munasinghe 1993). It is an essential bridge between the traditional techniques of decision making and the more environmentally sensitive approach now emerging. Environmental economics helps us to incorporate ecological concerns into the conventional framework of human society. Environmental degradation may be said to arise from human activity, leaving aside, for the moment, natural disasters and other events of human origin. Environmental economics plays a bridging role by mapping the environmental assessment results into the framework of conventional economic analysis. Once again, a variety of environmental economic techniques including economic valuation of environmental effects at the local or project level, integrated resource management at the sector or regional level, environmental macroeconomics and environmental accounting at the country wide, multisectoral level, and global or transnational environmental economics at the international level, facilitate this process of incorporating environmental issues into traditional decision making (Munasinghe *et al.*, 1993).

Natural Resource Management: The Changing Approach

A natural resource ecosystem is an integrated ecological system, one element of which is a product of direct or indirect use to man. The product may be biological as in the case of forests, agricultural products, fish and wildlife, physical in the case of water, air and soil or both. In all the cases, the distinguishing facet of a natural resource ecosystem is that man has a direct Involvement in the complex set of ecological interactions.

Developing nations are beginning to realize that natural resource degradation is a serious impediment to their central goals of economic development and the alleviation of poverty. Thus these poorer countries are seeking sustainable development paths that will permit them to simultaneously address both socio-economic and environmental concerns.

The identification of sustainable development options requires:

1. Good understanding of the physical, biological and social impacts of environmental degradation;
2. Precise estimate of the economic value of damage to the environment, to improve the design of policies and projects, and to arrive at environmentally sound investment decisions; and
3. Development of policy tool and strengthening of human resources and institutions to implement viable strategies and manage natural resources on a sustainable basis.

The concept of sustainable development encompasses three major points of view: economic, social and ecological. It is a formidable task to reconcile these various concepts and implement them as a means of achieving sustainable development. The social concept of sustainable development is people-oriented, and seeks to maintain the stability of social and cultural systems, including the reduction of destructive conflicts (U.N. Environmental Programme 1991). In the twentieth century, as the scale of economic activities has expanded, the adverse effects of those activities on natural resources have increasingly demanded more attention. This has led to a reactive approach to environmental damage, through increased cleanup and mitigation activities.

Objectives of the Study

With the conceptual background discusses above the present study has been undertaken to fulfil the following objectives.

(1) Ecologic—economic evaluation of renewable resources for sustainable development.
(2) To undertake a village level case study of the land and water resources and its management to illustrate constraints to development through micro level planning.

(3) To prepare comprehensive local area action plans for sustainable development.

A proper intervention strategy can only be worked out if the status of natural resources along with their spatial distribution is understood fully by the planners, the land owners and the users. Involvement of the local people in this process brings out relevant, and at the same time, genuine problems that affect productivity. In addition, it would generate not only a sense of participation among the local people but also a desire to improve their landuse practices. The traditional knowledge latent with the society also could be brought out and blended with the modern technology.

Resource Assessment of the Study Area: The Ecological Setting

Andoorkonam Panchayat situated about 20 km North of Thiruvananthapuram city in the Trivandrum district covers an area of 13.96 sq.km. and lies between 8°34′ to 8°38′ North latitudes and 76°50′ to 76°53′ East longitudes. The Andoorkonam Panchayat is bounded by Pothencodu in the North East, Kadinamkulam in the South West, Kazhakuttom in the South East and Mangalapuram in the North West. As per the 1991 census, the total population is 26,201 distributed in ten Wards. Physiographically the western portion of the panchayat falls within the lowland region comprising of mainly the coastal plain where as the eastern portion falls on the midland region with crests, side slopes and valley floors.

Land Resources

Land resources have been discussed in terms of landform, slope, relief and surface material. These are Ridge Crest, Hill Crest, Side slope-Moderate (5°-10°), Side Slope-Gentle (0°-5°) Valley Head, Valley Floor, Foot Slope, Sand Ridges and runnels.

Table 18.1 gives description of area under each of the land types estimated through map measurement. The sand ridges contributed the single largest landform covering 24 per cent of the area of the Panchayat.

Ridge and Hill Crest unit covers around 25 per cent of total area of the panchayat. This unit also includes part of the upper slope. It usually has a slope of less than 5°. These flat topped ridges are laterite and represent erosional surface. Side Slopes, the height of the laterite terrace is only 20 to 40 metres, the hill slopes are not too steep. Moderately steep side slopes with 10° to 15° contribute only 3 per cent of the slope area. There exists only few patches of scrap faces that suffer from acute soil erosion problem.

TABLE 18.1: Distribution of major landform unit, Andoorkonam panchayat

Landform unit	*Area (in sq.km)*	%
Ridge Crest	3.08	22.06
Hill Crest	0.03	0.21
S.S. Mod (5°-10°)	2.89	20.70
S.S. Gen (0°- 5°)	0.54	3.87
Valley Head	0.26	1.86
Valley Floor	2.64	18.91
Foot Slope	0.38	2.72
Sand Ridges	3.45	24.71
Runnels	0.62	4.44
Pond	0.06	0.43
Total	13.96	100.00

Source: Estimated from Asset Map of Andoorkonam Panchayat.

The valley floors constitute a narrow strip along the foot hills stretching from East to West in the North-Eastern part of the panchayat. It forks into two in the midway. The upper fork constitutes part of a micro watershed falling fully within the panchayat, while the lower fork constitutes a part of a larger watershed, major portion of which falls in the adjoining panchayat. The soil here is riverine alluvium deposited by the rivulets that flow through the valley. *Sandy Plain*, typical of Kerala coast is below 20 metres contour. The soil is coastal alluvium loamy sand type and is of low fertility due to the absence of organic-bio ingredients. Because of 1.5 metre clay layer below the surface, the soil is imperfectly drained. The plain is interspersed with sand ridges running parallel to the coastline creating shallow sandy basins in between. Runnels, the interdune low lying area are marked under this category. This type of landform is characterised by wetness with high proportion of silt/clay content. Normally it is given for paddy cultivation. However, widespread reclamation of runnels accommodating tree crops is a regular phenomenon.

Water Resources

An effective landuse plan can be taken up unless water resources are given due consideration. Like other parts of Kerala this Panchayat is also well endowed with rainfall. The rainfall pattern at Nedumangad weather station near to the panchayat gives the seasonality of and quantum of rainfall during the year 1975 to 1993. The average rainfall in the panchayat has been 153 cm for last 19 years. Around 42.91 per cent of it falls in the months of May, June and July. The second monsoon during October-November contributes 29.86 per cent of the rainfall. The heavy concentration of the rainfall within a short duration raises two sets of issues, (a) problem of drainage during monsoon season and (b) problems of water scarcity during summer. These problems are aggravated by the high run off from laterite hills.

Water resources of this area, or for that matter, any area can be described in terms of (1) surface water and (2) sub-surface water. In this case our discussion will mainly centre around the localised surface water run-off and sub-surface water because they are of utmost importance in micro level planning.

Surface Water

So far as the surface water is concerned, two hydrological regimes are noteworthy. These are (1) the drainage system in the coastal plain and (2) the drainage system of the laterite area. Attention may also be drawn to the natural drainage system for the coastal plain. Given the sand ridges that run parallel to the coastline and shallow basins, natural drainage during monsoon must have been towards the north western direction. It is not possible to clearly delineate these drainage flows due to widespread human intervention in the natural system. What is emphasised here is the importance of such a net work of drainage channels given flat terrain and imperfect soil drainage quality in order to prevent water logging. In addition it is to be noted here that water potential in this terrain is considerably high.

Five sub systems under the laterite area, of which two are major rivulets that originates from the valley head, and the other three are seasonal because they originates at very low altitudes and have small catchment areas. The rivulets that flow through the valley is mostly rainfed. This is particularly so for the rivulet in the upper fork of the valley which originates in the laterite terrace zone. The catchment area of this rivulet is relatively small and it dries up for most part of the summer. The second rivulet flowing through the lower fork of the valley along the panchayat boundary has a larger catchment area and originates from a head pond in the upper portion of the second valley. But even this rivulet would dry up in the height of summer. The valley rivers largely perform the function of drainage channels during monsoon and to a lesser extent function as irrigation source during post monsoon season.

Ponds and Tanks

Apart from the streams, another important surface water resource is provided by ponds and tanks. In fact, these have multiple use like minor irrigation, fish farming, domestic water use etc. About 23 ponds are recorded in the panchayat. The largest pond has an area of 0.06 sq.km. Out of these 23 ponds, 3 are seasonal and the rest 20 are perennial. Table 18.2 provides ward level distribution.

TABLE 18.2 : Wells and ponds in Andoorkonam panchayat

Ward No.	*Wells*		*Ponds*	
	Perennial	*Seasonal*	*Perennial*	*Seasonal*
I	316	40	1	1
II	370	0	4	1
III	44	45	1	0
IV	279	12	1	1
V	100	192	5	0
VI	162	37	0	0
VII	420	0	0	0
VIII	259	0	0	0
IX	428	1	0	0
X	477	7	8	0
Total	2855	334	20	3

Sub Surface Water

To understand the nature of the subsurface water condition, inventories have been carried out in almost all the open wells and ponds in the panchayat and the data recorded in a structured data format with reference to the well and pond location in the map on assets. There are 3189 wells in the panchayat. This gives a well density of 228 per sq.km, a significantly high density. Out of these 3189, only 334 wells are seasonal. This indicates a good water potential.

Finally, some remarks may also be made on the ground water availability. Only data on the depth to water table as measured from wells has been analyzed and therefore no rigorous statement can yet be made on ground water potential. Ground water is available in less than three metres depth throughout the coastal plain. There is also smaller patch of such area in the upper fork of the valley. In the rest of the valley zone depth to the water table is 3 to 6 metres. In most part of the laterite terrace the water table is between 9 to 12 metres of depth. In around less than a quarter of the laterite terrace, the water table is deeper than 12 metres from the surface.

A Profile of the Socio-Economic Characteristics of Andoorkonam

The persistence of economic backwardness of the State, despite improvement in human resource development, however, has emerged as a matter of serious concern. There is also a growing awareness of the non-sustainability of the process of economic growth in the State. The economic growth has been patterned on intensive utilisation of natural resources, many of which are nearly harvested such as forest, on the basis of unregulated cultivation in ecologically sensitive areas, such as hill slopes etc. Environmental degradation has been identified as a major impediment to agricultural growth. Kerala's achievement in key areas of human development particularly health and education have been widely appreciated. Illiteracy has been virtually eradicated from the State and infant mortality brought down nearly to the level of some of the developed countries in the world. The fertility rate has also seen a sharp decline.

Our data base for the analysis consists of the socio-economic census of the households in the panchayat conducted as a part of the Panchayat Level Resource Programme and 1991 census. The total land owned by the households as per the survey came to only around 78 per cent of the total land area of the panchayat. A part of the discrepancy could be explained by the village land owned by non-residents of the village and institutional land holdings such as of industrial units, hospitals and other public offices and Government *purampoke*, etc. Even if these factors are taken into consideration it is evident that there was a considerable under-reporting of the land owned by the households. It could possibly be that the respondents reported only holding directly under him/her. Despite this serious limitation of the data it is possible to delineate certain important characteristics of the village population.

Demographic Characteristics of Andoorkonam Panchayat

The work is mainly based on the primary data collected from the field survey during the Panchayat level Resource Mapping Programme and also the census data (1991) obtained from the National Informatics Centre, Kerala State Unit, Trivandrum for comparison.

Population Distribution

A major trend has been the growing demographic pressure on land. According to the 1991 census, the total population in Andoorkonam panchayat is 26,201 (137,83 males and 12,418 females). A major drawback of the socio-economic survey conducted in 1993 in Andoorkonam panchayat is that around 20 per cent household were not covered due to various reasons. Hindus, Muslim and Christians constitute the population of the panchayat. Hindus are in majority followed by Muslims and Christians.

Density Distribution

The density distribution has been worked out at the ward-level (Table 18.3). The population density of this panchayat is as high as 1877 persons per sq.km. The highest concentration is observed in the western part. As one moves towards the east, density decreases.

TABLE 18.3: Distribution of population, density and size of household of Andoorkonam (1991)

Ward No.	*Male*	*Female*	*Total*	*House holds*	*House hold size*	*Density*
I	1131	1135	2266	409	5.54	1069
II	1085	1093	2178	452	4.80	1462
III	927	958	1885	394	4.78	1533
IV	1090	1104	2194	429	5.11	1701
V	1198	1208	2406	464	5.18	1731
VI	3665	1949	5614	821	6.84	4420
VII	1104	1158	2262	472	4.79	2308
VIII	912	1025	1937	356	5.44	1309
IX	1338	1440	2778	518	5.36	2750
X	1333	1348	2681	493	5.44	1577
Total	13783	12418	26201	4808	5.45	1877

Source: National Informatics Centre, Trivandrum Unit, 1991.

Household Size

Analysing the data pertaining to average size of the households (number of persons per households) a very interesting picture emerges. The average size of household is 5.33 persons. The population of

Andoorkonam has been rising at a faster rate than the average for the State. Between 1981 and 1991, the population increased by 31.12 per cent.

Housing Conditions

The housing conditions of the households tend to bring out relatively better economic status of the households in Andoorkonam. The data on housing of particular households gives the possible under estimation in household income.

It is seen from Table 18.4 that 23.5 per cent of the houses are tiled and 17.24 per cent of concrete. Traditional thatching with palm leaves and grass have 10.7 per cent. 54 per cent of the houses were electrified and 67 per cent of the households had their own well or pond.

TABLE 18.4: House construction pattern—Andoorkonam panchayat

Ward No.	*Tatched Houses*	*Tiled Houses*	*Tiled cum Terraced*	*Terraced Houses*
I	105	78	140	76
II	89	104	87	34
III	5	93	45	34
IV	90	73	96	46
V	77	62	166	74
VI	29	176	201	36
VII	17	141	160	173
VIII	19	104	103	64
IX	43	122	141	125
X	41	177	113	167
Total	515	1130	1252	829

Distribution of Scheduled Caste, Tribe Population

The distribution of Scheduled Caste (SC) population does not indicate any specific trend (Table 18.5). This is yet another speciality of Kerala's demographic feature wherein the Scheduled Caste and Scheduled Tribe population are dispersed in almost all areas. Virtually the Scheduled Tribe population is nil in the panchayat.

Literacy

Kerala has attained the highest effective literacy rate in the country. The literacy rate achieved is 89.81 per cent compared to the all India rate of

52.21 per cent. Andoorkonam panchayat as a whole has a literacy rate of 78.19 per cent.Literacy rate in the area varies from 75.07 per cent in ward I to 83.29 per cent in ward VII.

TABLE 18.5: Scheduled caste and tribe population—Andoorkonam panchayat

Ward No.	*Scheduled Caste*		*Scheduled Tribe*		*Total SC & ST house holds*	*Total SC & ST popula-tion*	*House hold size*	*Total popula-tion*	*% to total popula-tion*
	M	*F*	*M*	*F*					
I	96	112	0	0	13	208	16.00	2266	9.18
II	129	125	0	0	32	254	7.98	2178	11.66
III	26	24	0	0	NA	50	NA	1885	2.65
IV	176	192	0	0	50	368	7.36	2194	16.77
V	271	245	0	0	46	516	11.22	2406	21.45
VI	939	922	9	0	323	1870	5.79	5614	33.31
VII	6	4	0	0	NA	10	NA	2262	0.44
VIII	178	206	0	0	31	384	12.39	1937	19.82
IX	281	292	0	1	40	574	14.35	2778	20.66
X	104	99	0	0	NA	203	NA	2681	7.57
Total	2206	2221	9	1	535	4437	8.29	26201	16.93

Source: N.I.C. Kerala State Unit & Panchayat Booklet.

Working population

Working population in this area covering main workers and marginal workers accounts for 36.87 per cent of the total population. *Agricultural workers:* The landuse pattern of the State shows that the land under cultivation constituted 57.9 per cent for the year 1992-93 of the total land area which was only marginally higher than the previous years level of 57.86 per cent (*Economic Review*, GOK 1994). The cropping pattern of the panchayat shows that the land under cultivation constitutes 97.84 per cent. The area under each crop is estimated from landuse map in which major crop in each plot had been marked. A plot which could not be identified with a major crop has been identified as mixed crops/ trees. Coconut is the most important crop accounting for 32.05 per cent of the area. Coconut is also an important component of the "mixed crops"and "mixed trees". Paddy today is grown only in 12.78 per cent of area.

Occupational Distribution

Totally, 7246 male and 2028 female workers are employed in different sectors in the panchayat (Table 18.6). Thus the labour participation (Main workers) rate in the panchayat is 35 per cent. The unemployment rate of the panchayat is around 50 per cent (excluding the age group of 0-6).

TABLE 18.6: Occupational structure of the households—Andoorkonam panchayat

Word No.	*Self-Employment*	*Salary/ Regular wage*	*Temporary or Daily wage*	*Income from abroad*	*Other Jobs*
I	17	55	288	39	0
II	95	19	200	0	0
III	50	30	66	29	0
IV	30	109	163	1	2
V	65	60	160	86	6
VI	23	106	293	20	0
VII	69	151	235	36	0
VIII	4	26	244	14	2
IX	261	86	73	10	0
X	18	137	226	110	7
Total	632	779	1948	345	17

Source: Socio-economic survey, 1993.

The alarming growth in unemployment continues to be a major socio-economic problem faced by the State of Kerala. As on May 31, 1994, there were 40.2 lakh job seekers on the live registers of the employment exchanges in the State. Out of this, the educated job seekers with qualification S S L C and above were 26.6 lakhs which constituted 66 per cent of the total job seekers (Table 18.7).

Ecological Peculiarities of Valleys

The valleys are fluvial in origin and the streams passing through the middle of the valleys can be identified as the third or fourth order member of a stream network usually starting from the highlands, reaching rivers which ultimately join the sea. Kerala is a high rainfall area with two monsoons—the Southwest and the Northeast. Because of the steep slopes and gravelly loam texture of the uplands, most of the water reaches the intervening valleys as surface run off or as subsurface flow. The fine grained silty-clayey material in the valleys imbibe the water and act as cushioning material. During rains, after saturation in the valleys, the water is flushed out through the streams which later reaches rivers and then to the lowlands, backwaters and ultimately to the ocean. This is the water control activity played by the valleys.

TABLE 18.7: Unemployed rate in each ward of Andoorkonam panchayat

Ward No.	*Unemployed*		*Total*		*Unemployed*	
	Male	*Female*	*Male*	*Female*	*Male %*	*Female %*
I	413	912	1131	1135	36.52	80.35
II	457	882	1085	1093	43.12	80.70
III	276	563	927	958	29.77	58.77
IV	387	749	1090	1104	35.50	67.84
V	437	859	1198	1208	36.48	71.11
VI	633	1057	3665	1949	17.27	54.23
VII	532	905	1104	1158	48.19	78.15
VIII	338	650	912	1025	37.06	63.41
IX	592	968	1338	1440	44.25	67.22
X	532	1020	1333	1348	39.91	75.67
Total	4597	8565	13783	12418	33.35	68.97

Source: N.I.C., Trivandrum.

But in recent times, due to the economic and institutional reasons, the valleys are used for a variety of other uses—agricultural and non-agricultural which need conversion of the valleys. For this, valleys are usually filled up and the ground is raised with coarser material like laterite and channels are made to drain off the standing water. All other uses like cultivation of coconut, banana or vegetables need only flowing water. Non-agricultural uses like building of houses, industries or clay mining do not need standing water at all. So the raising of ground by coarser surface material and channelling will lead to excessive run-off to lowlands leading to water-logging and flooding of these during rains. This deprives the surface and ground water reservoirs in midland areas their due share of rain water which leads to drought conditions during summer.

Environmental Appraisal for Desirable Landuse

The valleys are water faciled and the impermeable clay formation allows only minimum percolation loss. Due to the high rainfall and steepness of slopes of intervening upland, the valleys in Kerala exhibit the characters of a wetland eco-system. In Kerala, the highland, midland and lowland valleys are traditionally being used for paddy cultivation.

Agricultural Practices and Ecozone Wise Production of Some Selected Crops

An important reason for low per tree productivity of coconut is the practice of planting coconut too closely to each other. Density of coconut planting is an important factor that influences the per tree productivity. The optimal spacing of coconut would vary from one locality to another and also the pattern of planting. Coconut trees recommended by the Coconut Development Programme is 70 per acre on the average, so that inter crops may be grown.

There is a powerful economic rational to the persistence of dense planting practice despite perceptible decline in the tree productivity with the increase in density. In the absence of intercropping or inadequate labour and fertiliser and irrigation inputs, dense planting is rational methode to maximise the output per acre. Without intercropping and intense cultivation practices, optimal spacing that maximises nuts per tree may not maximise the total income of the farmer. Wrong agricultural practices are prevalent not only in the spacing of trees but also in planting methodes particularly in the reclaimed paddy land. The productivity of coconut trees in the water logged areas is relatively lower.

We are not taking up the economic factors such as the output and input prices as they are largely determined in the macro economy. However, the importance of a favourable macro economic environment must be emphasised. The recent increase in the fertilizer prices and decline in the price of coconut has already created much disquiet.

With non-agricultural activities as the major source of income there is very little compulsion from the land holders to maximise the agricultural production.

In paddy also there is preference for traditional varities and the use of HYV seeds has tend to decline over time. The traditional seeds, though low yielding, are more sturdy and versatile. With the decay of traditional water harvesting measures through siltation of the water channels and ponds in the valley, the risk to summer crop has increased. Similarly, flooding during the monsoon has become a threat to the first crop.

A major problem connected to the agrarian structure is the increased scarcity of agricultural labourers particularly during the peak season.

With the spread of education the younger generation in agricultural households have been withdrawing field labourers. This is particularly true of paddy cultivation.

Conversion of Paddy Fields to other Crops

Trend in Andoorkonam

Diversion of land from one crop to another is related to the socio-economic factors to a large extent. Increasing cost of cultivation, higher labour input, need for intensive care and more susceptibility to seasonal vagaries have together provoked the farmers to change their cultivation practices from paddy to coconut or other tree crops. It has been observed that in 1965 about 2.08 sq.km. of land was under paddy, which has come down to 1.83 sq.km. in 1993. A fall of 12 per cent in 28 years. The paddy fields in the coastal plain has been completely converted (Map 18.1). The valley areas are reclaimed along the slopes and in the head. Proximity to settlement is a major contributory factor in selecting the reclamation site.

Conversion of paddy field to other crops is widespread. A major reason for the shift to coconut from paddy is that it requires less personal attention from the land owners. The high wages and lack of commitments on the part of the agricultural labourers were common complaints. Paddy requires annual investment and the return is crucially depended upon rainfall. This uncertainty in contrast to the regularity of income from coconut land has been a major factor influencing the crop choices. This trend is observed in other parts of the State also. It will be interesting to present a State level picture here to demonstrate the gravity of the problem.

Socio-Economic Reasons Behind the Physical Changes

An attempt is made to bring out the socio-economic and institutional factors responsible at the micro level for the landuse changes, particularly paddy and coconut cultivation. To get the information of the evolution of paddy cultivation and problems at the lowest level, i.e., at the plot level, interviews were held with some farmers. An attempt is also made to calculate the cost of cultivation to get insights of the relative profitability.

Case studies of some of the farmers in the village summarised from all the zones have been chosen for the presentation.

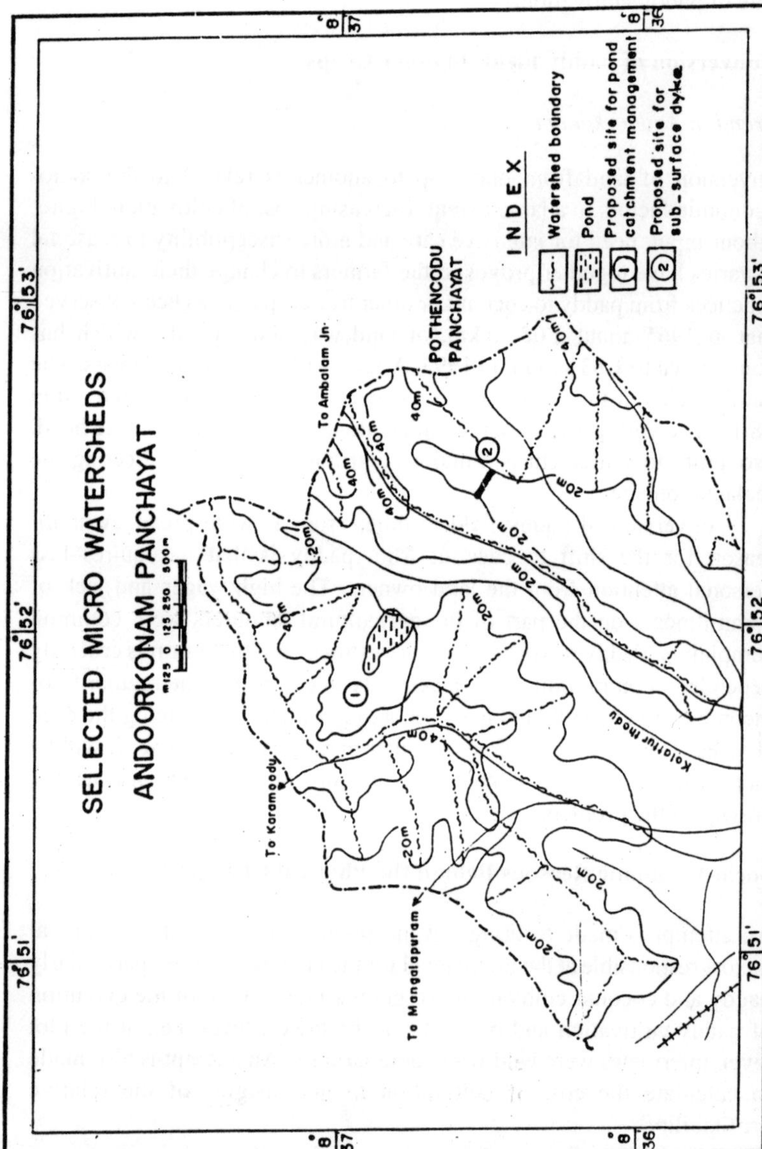

FIG. 18.1: Selected micro watersheds Andoorkonam panchayat

Case 1: One of the farmers cultivates 0.18 acre of paddy field in the valley with double crop practice. According to him, labour scarcity is the main problem and is very difficult to finish the transplanting in time. Also there are competition among farmers to get a hired tiller/oxes for ploughing the field. Average cost incurred by him for paddy production is around Rs.510/ and the average yield of paddy is 90 kg. *Case 2:* In 30 cents of coconut land 25 palms are planted with mixed trees in the Ridge crest zone of the panchayat. Cow dung is used for manuring. Average yield of 400 nuts per year was obtained. The age of the palms comes up to 25 years at the most. *Case 3:* In this zone a cultivator has planted 20 palms in his 30 cents of land. No inter-cropping is practiced . Seedling used are both HYV and traditional varieties. He is hopping that he is getting around 1600 nuts per year. Both traditional and chemical fertilizers are used for manuring. The high price of chemical fertilizers makes him to reduce its use. He is a successful farmer, through agriculture is not his primary employment. *Case 4:* One coconut garden of 16 cents in the slope zone side have 15 palm. Average yield of 1000 nuts per year was obtained. Cow dung and bio-waste were used for manuring.

Coconut cultivation is preferred compared to paddy and other crops because of the low annual inputs which are required for coconut. The data on the production, yield and area under paddy, collected from the "Krishi Bhavan" in the panchayat shows a decline in the production due to a decline in area from 1991 to 1995 despite a positive growth rate in yield. During 1991 the total area under paddy in the panchayat was around 150 hectares and in 1995 it comes to only 125 hectares and the production per hectare is 1300 to 2000 kg respectively. It may be noted here that the total area under paddy is 183 ha. as emerged through resource mapping. All these land are not put to paddy cultivation. Current fallow is also reported in this panchayat. Group Farming system was introduced in the panchayat with the support of the "Krishi Bhavan" for the paddy cultivation and HYV seeds were also distributed among the farmers. Average cost incurred is around Rs.3350 per ha and the average yield of 2000 kg is obtained per hectare (Table 18.8).

In a mixed cropping system, as in Kerala, competing crops have definitely affected the landuse decisions of farmers. The best example is the phenomenal expansion of rubber cultivation in recent years in the study area. Panchayat is having an area of 0.82 sq.km. of rubber plantation with the average yield of 150 kg per day/hectare. Rubber plantation does not require frequent and close attention that a seasonal crop like paddy requires and hence suitable for the part-time farmers.

TABLE 18.8 (A): Production and cost of cultivation of coconut (Ecozone-wise)

	Garden land	Zone *I*	*II*	*III*	*Entire Panchayat*
Tree (Nos.)	1 Acre	83	66	93	71
Production (No. of nuts)	1 Acre	1333	5333	6250	2336
Cost of Cult. (In Rupees)	1 Acre	750	900	800	600

TABLE 18.8 (B): Production and cost of paddy cultivation

	Paddy land	*Zone IV*	*Entire Panchayat*
Production	1 Acre	1000 kg.	1300 kg
Cost of Cultivation	1 Acre	2550 Rs.	1400 Rs.

A preliminary attempt to calculate the cost of cultivation and relative profitability showed trends unfavourable to paddy. The phenomenon of conversions are found to impose unidirectional externalities on the cultivating farmer which leads to more conversions.

Intervention Strategy for Sustainable Development

Sustainable development is essentially a planning processes with definite goals and priorities. The scale of operation is as important as the methodology itself. The "trickle down approach adopted in various planning activities has not produced desired results. Moreover there is also a serious lacuna in understanding the "complex ecological interdependence" resulting in lack of integrated approach covering land and soil; ground water and surface water; forestry, agriculture and pastoral development; urban and rural development etc. In the case of sustainable development, it is considered both pragmatic and rational to shift to the Local Level Planning—the panchayat, the lowest, administrative unit in India (Chattopadhyay , 1993).

It is a fact that environmental degradation and socio-economic problems are expressed mostly at the local level (Carpenter and Harper, 1989), Barbier (1987) pointed out the importance of regional and local variations and stressed that individual economic-environmental system may have to be assessed for sustainable development. As the problems are manifested at the lowest level it has to be tackled there itself.

Therefore, sustainable development calls for intervention at the various levels, i.e, ecologic, economic and institutional. This chapter is designed to highlight intervention strategy, to be taken up for the Andoorkonam Panchayat.

Land and Water Management in the Panchayat

Broadly the intervention programmes will be designed as follows:

1. Land and Water management.
   (a) Landuse (Agricultural) Management.
   (b) Watershed development programme.

The thrust of the action programme in the primary sector is to develop an integrated approach to land, water and crop management. The programme aims at increasing the productivity of the various crops and generating employment through setting up of a series of agro-based service and production enterprises.

It is now widely recognised that participatory development is the key to sustainable and environmentally sound development. The underlying philosophy of the programme is (a) to adopt an approach to understand the nature of complex interconnections in the human-land-water-biomass system; (b) to evolve a comprehensive ecodevelopment action programme in order to continuously enrich the health of the resource base consisting of soil, water, vegetation, livestock, fisheries, genetic diversity etc.; and (c) to consciously arrive at a strategy of sustainable resource use based on (a) and (b) above.

The integrated land, water and crop management could involve the following steps:

1. Investigation of resource dynamics and documentation of traditions of resource use, folk knowledge etc.
2. Identification of socio-ecological causes of resource degradation and drawing out and implementing eco-development action packages for corrective and preventive intervention.
3. Implementing new packages for productivity enhancement and for value addition conjunctively with the eco-development packages designed already.

The disquieting features of agricultural performance prompted us to pursue the micro level analysis of constraints to agricultural growth. Given the ecological diversity, differences in cultivation practices, social traditions and characteristics of farm households, it is important to analyze specific factors at local level. The panchayat of Andoorkonam in Trivandrum district was the case study chosen. Even a small village like Andoorkonam, on a close analysis, reveal a surprising degree of ecological diversity, from hill crest, slopes, valleys, to sandy plains (Chapter III). Even in micro level intervention to improve agricultural productivity, these differences have to be borne in mind. There is a high level of fragmentation of holdings and occupational shift away from agriculture due to which agriculture has already become a subsidiary occupation. In such a social environment, how could so sufficient social enthusiasm be generated for agricultural rejuvenation, or appropriate social institutions be created for better utilization of land and water resources.

In this stagnant situation it is certain that poverty cannot be eradicated by ad hoc development activities, as has been attempted to in the past. For an enduring solution, additional incomes will have to be generated in the village itself and the only way to accomplish it, is to enable the poor to take up supplementary enterprises in the fields of horticulture, agriculture, animal husbandry, poultry keeping and household industries. Each of these can give higher returns than that is possible to get from conventional farming on small and dry farming areas. With this end in view, farmers should be encouraged to select one cash crop or enterprise most suited to the person and the area. Then all the necessary assistance such as technological, financial and managerial, should be provided through the efforts of panchayat authorities. To promote such specific programmes, it is essential to provide participative extension strategy by the various departmental agencies.

The two important crops in the panchayat are coconut and paddy, the later being cropped twice (P1 & P2) a year. A recent trend that is observed among the cultivators is a marked decline in the interest in paddy cultivation, as is mostly the case throughout the State. This has resulted in the decrease of area coming under paddy cultivation, viz, about 1.83 km^2 More irrigation facilities can be created by properly maintaining and deepening of the ponds, digging of borewells which are not having sufficient water, construction of few more tanks in proper places, digging of some more ponds, if necessary, and by extensive adoption of well irrigation. These activities can create additional man days of labour. They can also help to increase the productivity of paddy.

Further, these increased irrigation facilities would provide ample scope for economies of large scale returns by augmenting the productivity of the nearby perennial crops and would help in drinking water supply to the area. The blocked water generates water source in dry land wells. The main crop in the unirrigated dry land area in the panchayat is coconut. The low productivity of coconut can be increased, if proper irrigation facilities are provided along with planting of high yielding varieties of trees. Apart from high yielding (HYV) of seeds, fertilisers and implements, a major input for creation of more employment and income in the panchayat is credit or finance, especially for those below the poverty line. For all these developmental activities, the farmers should be financially assisted by the panchayat/co-operative sector with easy credit at low interest rate. If these activities are initiated and successfully implemented by the panchayat, then it is definite that the production of the two major crops will increase.

The Landuse Practiced and Desirable in Andoorkonam

The landform map of Andoorkonam showed that the slopes have moderate and gentle, laterite soil, poor water availability. So a dry crop such as cashewnut is the best land use for this area. The midslopes are steeper, water availability and soil thickness are moderate. There is limited moisture retention here and hence, so this is ideal for tree crops like mango, jackfruit etc. and plantations such as rubber. The lower slopes have moderate to good water availability, moisture retention, soil depth and there is scope for leached in nutrients from the upper parts. Here crop like coconut, vegetables, tapioca, banana etc. can be grown and the valleys can be devoted for paddy, vegetables and legumes in rotation. But we have seen the deviations from the practiced to the desirable and phenomena like the unbalanced expansion of area. This effects the landuse decisions at the local level in an unbalanced way, the consequences of which is beginning to being felt, especially in the midland terrain.

Watershed Development Programme in Andoorkonam

The programme aims at increasing the productivity of the various crops through judicious land and water management and generating employment through setting up of a series of agro based service and production enterprises. It is now widely recognised that participatory integrated development of watersheds is the key to sustainable and environmentally sound development.

TABLE 18.9: Ecozone wise—suggested landuse

Ecozone	*Suggested Landuse*
Ridge Crest	Mixed trees with basal cover, settlements, cashew, rubber (Limited)
Hill Crest	do
Side Slope Mod. (5°-10°)	Mixed tree crops, coconut, settlement, Terraced homestead, wherever necessary
Side Slope Gen (0°-5°)	Mixed tree crops, vegetables, settlements, tapioca
Valley Head	Vegetables, arecanut, banana, coconut
Valley Floor	Paddy, vegetables & pulses in rotation, not to be reclaimed
Foot Slope	Mixed trees, tapioca, settlement, vegetation
Sand Ridges	Coconut, settlement, cashew,vegetation
Runnels	Paddy, vegetables, banana, not to be filled up

The watershed of Andoorkonam stream can be divided in to A1 and A2 based on the topographical features of the area. The streams are originating from the lateritic sideslopes and is formed by two major 2nd order streams. The points of intervention are marked in the Map 18.1, pond management is considered to be an important programme. A sub surface dyke has been suggests for moisture retention at the upstream area. These two are just demonstrative samples. A number of sites can be worked out at the ground level.

The watershed programme shall be implemented through a Watershed Management Committee (WMC). The committee shall consists of elected ward members from all the wards in all the panchayats forming part of the watershed, retired technical personnel in the panchayat, representatives of farmers, landless agricultural labourers, unemployed youth, women and representatives of various line departments.

1. Increasing the productivity and production of paddy and other crops in the watershed area.
2. Revitalizing the stream network, and
3. Strengthening of pond-tank network. The ponds and tanks can provide additional water for irrigation for other crops and also can be used for agriculture.

Concluding Remarks

This micro level study covering ecological and socio-economic aspects of the Panchayat has brought out a number of interesting features. While it is not claimed that this study is exhaustive, it certainly indicates the dimension of various issues operative at the micro level. Sustainable developments in operational terms is primarily concerned about farm level production management. In that context, this study provides micro level frame work within which the production issues have to be dealt. It has been brought out through this study that there is a significant contribution of the social institutions in the emergence of micro-environmental problems like water logging, chocking up of drainage channels, reclamation of low land, under utilisation of pond etc.

It is now well accepted that the social, institutional issues have to be tackled by the local organisations with peoples effective participation. The various intervention sectors identified in this concluding chapter will provide the planning frame for the panchayats.

Apart from its importance in micro level planning the issues dealt with in this thesis have macro level dimensions. Therefore this study has to be viewed in two different levels; at one level it is a guide line of the Panchayat Planning for sustainable development and at the other level it brings out the issues which are micro level manifestation of macro level decisions.

REFERENCES

Barbier, E., 1989. *Economics, Natural Resource Scarcity and Development: Conventional and Alternative Views*, Earthscan, London.

Bernett. H.J. and Morse, C., 1963. *Scarcity and Economic Growth: The Economics of Natural Resource Availability*, Johns Hopkins Press, Baltimore.

Bureau of Economics and Statistics, Kerala, 1983-84. Seasons and Crop Report of Kerala State.

Carpenter, R.A. and Harper, D.E., 1989. Towards a Science of Upland Management in Developing Countries, *Environment Management*.

Census of India, 1991. Final Population Totals, Series 12, Paper 3, Director of Census Operations, Kerala.

Centre for Earth Science Studies, 1991. Panchayat Level Resource Mapping—An Approach Paper, A model for Micro level resource survey with people's participation, Thiruvananthapuram.

Centre for Earth Science Studies, 1994. Project Report on "Panchayat Level Resource Mapping for Decentralised Planning with People's Participation."

Government of Kerala, *Economic Review* various years, State Planning Board Thiruvananthapuram.

Mohan Munasinghe and Emst Lutz, 1991. Environmental—Economic Evaluation of Projects and Policies for Sustainable Development, Environmental Working Paper No. 42, The World Bank, Washington, DC.

Repetto, Robert, *et al.*, 1989. *Wasting Assets: Natural Resources in the National Income Accounts*, World Resource Institute, Washington, DC.

Srikumar Chattopadhyay, 1993. Sustainable Development: Concept and Application Case of Developing Countries, Indo British Seminar on Environment and Development, Lal Bahadur Shastri National Academy of Administration, Mussoori.

World Commission on Environment and Development (WCED), 1987. *Our Common Future*, Oxford University Press, New Delhi.

19

NEED FOR REVIVING COMMON-GOOD TRADITIONS AND AESTHETIC VALUES FOR SUSTAINABLE DEVELOPMENT
A Study in Southern Karnataka

B. ESWARAPPA

Introduction

A title of this kind generally leads the geographers including Social Geographers to raise their eyebrows. Though one wonder how this topic contributes to Geographical knowledge. In these days of interdisciplinary study of a phenomenon, nothing of this kind is a topic out of context. Certain aspects in human civilization repeat in course of time but they rarely get renewed. In a section of the society, loss of traditions and values not only have ramification on the people's living but also on the environment. Colonial rule, industrialization, polarized policies of Indian Planning efforts since 1950's, improvement in transport and communication have brought enormous, often negative, changes in our society. Now, we are in the cross roads of development and environmental degradation. Juxtapositioning this Indian rural scene, phenomenon of pseudo-urban biased environmental awakening phase is also in run. Increased urban rural contacts, inevitable dependency, too much exposure of excited ruralities are forgetting their common good traditions. Availability of mass produced building materials, knickknacks, plastics too eroding the aesthetics sense in the Indian rural scene. Loss of common good traditions with unsystematic popular schemes of present day governments have serious negative impacts on agricultural productivity and health of the rural mass. They are becoming highly dependents on inanimate energy, their tastes are turning towards urban mass consumption.

The Study Area

Though the title implies comparatively a maso region, about ten villages under Madhugiri and Koratagere taluk as of Tumkur District in South Eastern parts of Karnataka have been considered for the study. It is a region with rainfed crops like groundnuts, Ragi and minor millets. But tanks, open wells and late bore wells (tube wells with 50 m—120 m deep, 3.5 to 10. h.p.) are providing irrigation to grow crops like paddy, irrigated groundnut, mulberry and arcanut like garden cash crops. This part of Karnataka is not that rich in agriculture and mineral resources. Recent development of agriculture, particularly, mulberry cultivation with the help of bore wells irrigation and cultivation of groundnuts are becoming common feature. The dairy has been developed in most of the villages.

Introduction of mulberry cultivation (rearing of silk worm) and dairy also brought truck, bus and mini-lorry services to the most of the villages in this area. All the villages have daily bus services to Taluk, District and State headquarters like Madhugiri, Koratagere, Tumkur and Bangalore City, respectively. People can reach Tq. Hq. in about less than ½ hour; 1 to 1¼ hours to Tumkur and 3 to 4 hours at the most to reach Bangalore city by bus.

Statement of the Problem

Common good traditions which form the focal theme of this study refer to what a few decades ago (1950 and earlier), most of these villages used to have like desilting of tanks, reconditioning, digging, straitening of channels from a spring, repairing of approach roads from the village upto main road, concern for temple building repair with their own efforts etc. They were doing these under the work ethics of 'bitti' meaning 'free' service for the village.

As a matter of aesthetic values in these villages people were regularly holding Kolata, Rangoli, folk theater activities in the winter months of December, January and February when there used to be little work and season is dry having nights were clear sky. These activities,like many other things in our environment, are becoming endangered. Over the years people have acquired radios, T.V's. People are busy in growing multiple crops; finds little time for these aesthetic aspects of life. TV and its urban advertisements, people's exposure to urban way of life and education to a great extent have led to the impersonalization of village life and neglect of the common good traditions.

Objectives of the Study

Present study aims at the objective assessment of loss of some of the valuable common good traditions and aesthetic values in the selected ten villages of Tumkur district in southern Karnataka. The selected villages are characterised by fairly good wells and tanks irrigation and resultant mulberry cultivation. However, this need not necessarily mean that this study aims to assess agricultural development versus loss of these values. It is true that the loss of common good traditions is a general phenomenon taking place at relatively agriculture prosperous villages. The point is that comparatively prosperous people must initiate and try to retain these traditions. In the conclusion it is agreed that, the prosperous families of the villages must help the organizations concerned in protecting the rural heritage, values and environment. These prosperous people can work for the restarting are inculcation of these values compared to highly poor villagers.

Methodology

Keeping the objectives in view primary data have been collected in these ten villages. The aspects coming under rural economy are: area under irrigation, bore wells drilled since 1980, area under sericulture, milk production and acquisition of radio, T.V., and tractors. There is a perceptible growth in these aspects of rural economy and infact most of them have been newly introduced in to the rural areas since 1980. These variables are amenable to simple quantification and comprehension.

In order to understand the status of rural common good traditions and aesthetic values questions were asked to villagers on the basis of questionnaire and also information were collected by interview method. Personal observation by the author in these villages about these aspects go a long way in portraying the status of those aspects like desilting of tanks, channel repair (from tanks, stream and from the head of a spring), people's voluntary participation in other rural public utility activities like maintenance of temples and its Dharmashalas, repair of temple steps etc. People's participation in folk art, females' participation in folk art and use of herbs though observations are recorded in the questionnaire, the out come—as the social issues—are not truly amenable to quantification.

Aspects of Village Economy

Among the villages, selected for the study Shambonahalli, Raghuvananhalli, Ganjulagunte, Maruvekere, Halehatti, Chinakavajra, Siddapura, Bijavara are in Madhugiri Taluk of Tumkur district. Thumbadi and Koratagere lie almost adjacent to Koratagere taluk of the same district.

An effort has been made to bring out the development in agriculture by expressing the percentage of area under irrigation from tanks, open wells and recently drilled bore wells which provide common form of irrigation though it is not assured every year. Since the villages lie in the rain shadow region of south India, people heavily depend on rain, tanks and wells. Off late they are going for costly bore well irrigation (Table 19.1).

Along with bore wells, since 1980, sericulture has been introduced into these villages. With sericulture, people have also taken up milk production. There are milk co-operatives. Infact, introduction of sericulture and milk production has considered by many to have reduced migration to urban centres by providing employment in the villages almost throughout the year. Thumbadi, Mallekaveree, Bijavara, Chinakavajra, Maruvekere have sizable areas under sericulture and have good milk production. The relative position of these villages under these aspects are discernable from the Table 19.1.

This agricultural development has brought noticeable change in the construction of new houses, acquisition of radios, T.V. sets and in dresses too. All most all of these villages have daily bus, tempo, lorry services to carry men and materials. There is an increased mobility in the villages. It has been observed that, introduction of bore well irrigation and the development of sericulture in these villages has brought high dependency, mobility of villagers with the nearby towns and cities.

Villagers' Participation in Common Good Traditions

The author has observed that, during 1950's and earlier, all these villages had many common good traditions. They used to involve the whole of rural community. This indirectly promoted the conservation and sustainability of rural resources. One of the activities was desilting of tanks. Earlier people were desilting their tanks and using the silt soil for dry lands. It used to enhance the fertility and moisture retaining capacity of the soil in this area of uncertain rainfall. In fact, it is a great loss, unfortunately, people are going for chemical manures for dry lands,

TABLE 19.1: Salient economic aspects of the villages 1991-'92

Village	*Total Population (1991) as per village acct.*	*Total No. house-holds*	*Total Area (in hect-ares/Acres)*	*% of Irriga-ted Areas*	*Borewells (since 1980)*	*Area under Sericulture (in Ha)*	*Milk production (Ltrs. per day)*	*Radios*	*No. of Tractors*	*T.V. sets*
Shambhonahally	533 (8)	145 (8)	232.85 (7)	53.2 (5)	5 (8)	6 (8)	60 (9)	23	nil	6
Raghuvanahally	507 (10)	60 (10)	119.07 (10)	48.6 (7)	6 (7.5)	3 (9)	45 (10)	42	1	2
Ganjalagunte	2,471 (3)	220 (6)	205.87 (8)	61.2 (4)	21 (4)	14 (5.5)	85 (7)	112	3	8
Maruvekere	1,440 (7)	211 (7)	203.32 (9)	53.4 (5)	23 (3)	36 (2)	390 (2)	180	4	26
Halehatti	520 (9)	130 (9)	272.72 (5)	30.0 (8)	6 (7.5)	12 (7)	66 (8)	26	nil	1
Chinakavajra	2,749 (2)	527 (1)	2,124.50 (1)	16.6 (10)	17 (6)	14 (5.5)	250 (5)	60	1	10
Siddapura	2,846 (1)	414 (3)	318.62 (3)	66.0 (2)	20 (5)	13 (6)	380 (3)	234	5	30
Bijavara	2,037 (4)	334 (5)	297.34 (4)	64.5 (3)	10 (7)	30 (3)	280 (4)	186	3	25
Thumbadi	1,972 (5)	520 (2)	249.59 (6)	79.4 (1)	40 (2)	126 (1)	560 (1)	230 (6)	6	20
Mallekavu	1,556 (6)	340 (4)	416.37 (2)	28.6 (9)	55 (1)	16 (4)	230 (6)	212	1	11

Source: Data compiled from field work for the above villages in 1992.
(R) = Rank.

TABLE 19.2: Villagers' participation in common good activities 1991-'92

Village	*Tank desilting*	*Channel repair*	*Voluntary Labour*	*Rural Public utility activities*	*Folk arts*	*Use of herbs*	*Female participation in folk arts*
Shambonahally	Nil	Less	Nil	Nil	V. less	Less	V. less
Raghuvanahally	Nil	Nil	Nil	Nil	V. less	Less	Less
Ganjalagunte	Nil	Nil	Nil	Nil	Moderate	V. less	Moderate
Maruvekere	Nil	Nil	Nil	V. less	V. less	V. less	Less
Halehatti	Nil	Nil	V. less	Nil	V. less	V. less	Moderate
Chinakavarja	Nil	Nil	Nil	Nil	Moderate	Less	Moderate
Siddapura	Nil	Less	Nil	Nil	V. less	V. less	Moderate
Bijavara	Nil	Less	Nil	Nil	V. less	Less	Moderate
Thumbadi	Nil	Less	Nil	Nil	V. less	V. less	Less
Mallekavu	Nil	Nil	Nil	Nil	V. less	Good	Less

(Very Less, Less, Moderate, Good)

Sources: Field observation 1992. Data collected through questionnaire.

which is a failure in these areas. People have almost given up, this age old practice which is evident from Table 19.2.

People of these villages used to repair natural water courses, mend the channels annually. These may be from tanks and a spring head to the fields. But, this important tradition of natural water supply to the village lands with community participation is almost given up. This may be attributed to the free supply of electricity to the farmers. It has resulted in undue dependence on electric pumps and neglecting the use of water from these sources. It needs serious consideration.

Another area where people's participation is declining rapidly is in people's voluntary participation in Rural Public Utility services and activities like maintenance of temples, its way steps and Dharmashalas, construction of a small stretch of approach roads from the village to the main road. Earlier, these activities were carried out by village community under the concept of "Bitti" which means community free labour. In the recent years fast transportation false assurances by local leaders and the introduction of a sort of tourism in these areas have lead to the decline of people's participation in the protection of rural common good practices and properties. Table 19.2 in a way tries to measure this aspect of loss.

People of a few villages still have some form of folk art. But in most of the villages the original folk art like classical play, folk songs during car festivals etc. are not performed in the earlier enthusiasm. Growing population and its pressure on limited land resources, urban contacts, introduction of radio, off late T.V. sets into villages have caused immense damage to the folk art in these areas. A day has come that a Journalist or a T.V. man comes and tries to revitalize these to some extent in the villages.

Till very recently females in these villages used to perform moon light rangoli ritual annually. It was a beautiful community participation. Women used to draw beautiful designs on the ground singing till late night. It was strain easing ritual for poor souls. Increasing scarcity of labour, radios and very recently inroad of T.V. have again almost put a full stop to this valuable tradition. One of the out comes of this study is the author is going to revive this folk art in his village which is now nonexistent in his village.

One of the great losses of our tradition, particularly, in these villages is the fast declining trend of lesser and lesser use of traditional herbs. When we, on one side, in urban areas through media advertisements are trying to use herbs; paradoxically our village folks are increasingly turning towards the harmful synthetic goods! Though we are not in a position to give up allopathic medicine, there are plenty of events to revitalize the use of our traditional herbs. In the current study an effort has been done to understand the use of herbs in these villages. Except Mallakavu, the story is same in all other villages. Since, this lone village lies little interior and it is closer to a hill "Siddara betta" a reservoir of herbs, there is a moderate use of herbs.

Similarly the loss of some management traditions have lead to the declining soil fertility, increasing cost of cultivation, falling yields, increasing dependency on inanimate energy, increased use of synthetic factory made mass consumer products with falling general health status. Though there is a rise in agricultural incomes but the urban dependency is also on its rise. There is a wastage of energy, rural man power and loss of common good traditions in the villages.

Conclusions

It is such a paradox that our villagers are to be remembered about their past great cultural values and aesthetics by urban voluntary organisations or such other concerned organisations. However, their revival inculcation and practice will bring life, lyrics above all the community feeling which is under peril at present. Indirectly, periodic desilting of village irrigation tanks will have positive effects on agriculture. It increases water capacity of the tanks; desilted soil becomes a valuable mixture soil for red soils and helps dry crops cultivation. Instead of destroying 'Kattes and Kuntes' (very small tanks meant for drinking water for animals etc.) for agricultural land the villagers must strengthen and rebuild them. This will enhance the soil water recharge and provide water for animals and community washing. It is observed that, the winter night formal ritual—Beladingala Rangoli, with folk songs is almost dying and it must be revived. Present invasion of radio and T.V. have made village folks, women particularly individualistic dry and gossip monging in their leisure. Its revival has multifarious benefit for women. It leads to culture revival, mental relaxation and they get aesthetic exposure while drawing rangoli design in rice flour with different colours for hours with songs. It is also observed that the maintenance of old temples, the small 'mantaps' (dwelling of people in car festival time near temples) and steps to deities on hill tops are either neglected or interfered by the local government leading to community nonparticipation. In fact, these temples and all these common properties were well maintained in the early days by our elders with limited resources of course with their enormous sense of common good and concern for aesthetics in life. It is our duty to revive them, whatever little way and manner in our rural settings. So that, both of our rural cultural and physical resources conserved for the posterity. Infact, community feeling is required to promote village level community nonconventional energy, social forestry etc. Their success lies in the promotion of common good traditions in our vast rural society.

REFERENCES

Clark, R., 1975. *Notes for the Future—An Alternative History of the Past Decade*, Thames and Hudson Ltd., London.

Government of India, 1981. Census Handbook, Tumkur district.

Palhiwala, N.A., 1991. *We the People—India—The Largest Democracy*, Standard Book Stall, Bombay.

Phil, H., 1973. *Planning and Human Need*, Praeger Publishers, New York.

20

ENVIRONMENTAL IMPACT OF MINING
The Kerala Scenario

K. SOMAN

Introduction

MINERALS are an important part of the natural resources, and their increased use for production of metals and their alloys, fertilizers, chemicals, cement and power has in recent years revolutionised the industrial and agricultural base of many nations. Consequently, mining of mineral resources got an accelerated pace throughout the world.

As minerals are "geological ore concentrations of such magnitude that can be technically and economically extracted", the frequency of their occurrence and quality depend on geological conditions. Mining projects are frequently located in remote or underdeveloped regions of developing countries. These areas are often ecologically and climatologically very sensitive, without adequate physical or social infrastructure to guarantee environmentally sound operations.

Mining projects alter the natural environment due to the magnitude, time frame and the nature of the mining process. The primary goal of metal mining is the extraction of non-renewable natural resources and this necessarily implies permanent environmental change, regardless of the technology or methods employed, with long-term significance to the overall well-being of the population.

It is thus clear that mining leads to environmental damage, while economic development and self-reliance call for increased mining activities of the available mineral resources. Though there is no alternative to the site of the mining operations, options as to the location and technology of processing can really minimize the damage to the environment.

As far as Kerala is concerned, the role of mining operations is very insignificant compared to the rest of the country, as amplified in Table 20.1, and hence the damage to the environment due to mining is also limited. However, mining of the available mineral resources of Kerala, especially the clays for brick making, is causing serious damage to the environment and adversely affecting paddy cultivation. In order to assess the damage to environment due to mining in Kerala it is necessary to have an idea about the mineral resources of Kerala.

TABLE 20.1: Production figures for some of the ores/minerals of Kerala as compared to the all-India production

Sl. No.	*Ore/Mineral*	*India (tonnes)*	*Kerala (tonnes)*
1.	Kaolin	745,162	59,312
2.	Ball clay	242,394	516
3.	Fire clay	663,483	4,571
4.	Limestone*	49,083,000	nil
5.	Sillimanite	17,129	7,341
6.	Graphite	33,848	nil
7.	Ilmenite	174,823	106,885
8.	Rutile	8,936	7,391
9.	Iron ores	44,090,000	nil
10.	Limeshell	158,119	82,669
11.	Mica	4,942	nil
12.	Silica sand	924,252	59,079

Source: 1. Indian Mineral Year Book, 1986, Vol. I & II.

* In 1987 Kerala's production of limestone from the Walayar quarry reached about 2 lac tonnes (Some Environmental Impact Studies on Limestone Mining in Walayar area, Palghat District, Kerala, I year progress report, 1987).

Mineral Resources of Kerala

The known mineral resources of Kerala include (1) placer mineral deposit of ilmenite-zircon-monazite-garnet-sillimanite at Chavara and the adjoining coastal tract, (2) glass sand deposit at Shertallai, (3) clay deposits (Kaolins, ball clays, fire clays and tile clays) found throughout the state (4) limestone, (5) lime shell, (6) iron ores, (7) gold, (8) graphite, and (9) chrysoberyl. Building materials (granites, charnockites, gneisses laterite, tile clays and sand) are excavated in large quantities.

Out of the known mineral resources of Kerala, beach placers, clays, limestone, lime shell and building materials are mined extensively. Graphite was mined from Vellanad during the period from 1898 to

1912, and it then constituted the entire Indian production of graphite. Likewise, old gold workings made by a mining people known as Karumbars exist along outcrops of quartz reefs in SE Wayanad, and gold placer occurrences are worked locally in the Nilambur valley. Chrysoberyl is recovered from pegmatites and alluvial gravels of Thiruvananthapuram district. Abandoned mica mines exist near Punalur.

Ilmenite-Zircon-Monazite-Garnet-Sillimanite-Placer Deposit

The major deposit in Kerala, and the richest in the country, is in the Neendakara—Kayamkulam belt. Major heavy mineral concentrations are located at Neendakara, Chavara and Kozhithottam. The deposit extends for 22.5 km with an average thickness of 10 m. Heavy mineral concentration of the Chavara deposit is more than 80 per cent. Concentration of ilmenite here varies from 65 to 78 per cent, rutile 4 to 6 per cent, zircon 6 to 9 per cent, monazite 0.4 to 1.5 per cent, sillimanite 5.5 to 7.5 per cent, garnet 0.3 to 0.4 per cent, quartz and other minerals 2 to 13.1 per cent. The placer minerals of the Chavara deposit are thought to have originated from the khondalite migmatite complex of southern Kerala during late Proterozoic times, while their transportation and deposition at the present sites are very recent geologic events. Monazite from the Chavara deposit contains 8.73 to 12.0 per cent ThO_2.

Glass Sand Deposit

White glass sand occurrences are reported from the western side of the Vembanad lake around Alleppey, Shertallai and Panavalli. The state department of Mining and Geology has estimated the resources at 70 million tonnes with silica content of about 99.4 per cent.

Clays

Clays of sedimentary and residual origins occur throughout the State. Kaolins are the most abundant of these resources and Kerala's resources of kaolin are the largest in the country. Substantial reserves of fire clays, ball clays and tile clays are also known to occur in different parts of the state.

Weathering of various rocks types outcropping in Kerala has resulted in the formation of residual clay deposits. Some of the good quality kaolin clays belong to this variety. Clay mines at Kundara and Pazhayangadi are the classical examples of this variety of kaolins.

Cyclic deposition of the weathering products in shallow water environments has resulted in the formation of the coastal sedimentary strata of Kerala and these are known for the presence of extensive kaolin beds, fire-clays and in some areas ball clays.

Limestone

Occurrences of crystalline limestone associated with the metasedimentary rocks is known from a number of places in Kerala. The biggest such occurrence is reported from Pandarettu near Walayar in Palghat district.

The Limestone deposit at Pandarettu occurs as an E-W trending isoclinal antiform with steep plunge within calc-granu-lites and associated garnet-biotite-sillimanite-graphite gneiss. Mineral resources are estimated at 23 million tonnes (Sivadas and Rao, 1975).

Limeshell

A chain of back water lagoons, enclosed on all sides by land except for narrow inlets of sea, dotting the Kerala coast form excellent areas for the growth of marine fauna such as lamellibranchs, gastropods, brachiopods, oysters etc. These organisms after death sink into the bottom of the lagoons and accumulate as beds of limeshell deposits often attaining thickness of 6 to 8 m. Important among such limeshell deposits are located at Padappakkara, Monroe island, Kavanad, Manakkadavu, and Neendakara in Quilon district; Pallippuram, Panavally, Thanneermukkom, Kainakary, Mattathilbhagom, Kumarakom, Vaikom, Kulasekharamangalam, Chembamuri etc. constituting portions of Vembanad lake; Kodungallur, Chavakkad, Karanchira etc. in Trichur district; Ponnani (Malappuram district), Quilandy, Beypore, Kadalundi and Elathur in Calicut district and Parappuram, Melur, Tallicherry, Baliapatom, Payyannur, Cheruvathur, Nileshwar, Kanhangad, Kasargode etc. in parts of Cannanore and Kasargod districts.

Building Materials

Precambrian gneisses, charnockites and granites occupying the highland and midland regions of Kerala are good sources of construction/building materials. Laterites occurring ubiquitously in the Kerala region are also used extensively as a construction material.River sand and clays from paddy fields and river banks also constitute important construction materials.

Mining in Kerala

All the known mineral resources of Kerala, except the beach placers, china clays and limestone, are not economically feasible for mining at present. As such, mining activities have mostly been confined to the above mentioning minerals, although records of graphite, phlogopite-mica and gold mining are known in certain parts of the state. Further, brick clays and river sand are mined quite extensively in Kerala. Laterite and hard-rocks (gneisses, charnockite and granite) are excavated for building purposes.

On a close examination of mining in Kerala, it is found that mechanised and sufficiently large scale mining operations are confined to the limestone quarry at Walayar and at the beach placer deposit of Chavara.

Mining of Beach Placer Deposit

Two major companies namely M/s Indian Rare Earths Ltd., and the Kerala Minerals and Metals Ltd. mine and process the beach placer deposit at Chavara and adjoining areas.

Beach placer mining at the Indian Rare Earths Ltd., Chavara, is carried out by buldozing the top beach layer from the surf zone upto the beach (about 5 m), and by pitting upto 1m feet depth in the beach. This material is fed into the plant and the light fraction waste material is dumped away in the sea. The excavated material, as a rule, is replenished soon by material brought through wave action. However, buldozing from the surf zone generally pushes the sea line inward. The extent of mining operations towards the interior does not generally exceed 5 m and therefore, interference with human settlements is limited and damage to vegetation is insignificant. Mineralised zones beyond this limit towards the interior are left un-mined.The sea-mining operations involve scooping of material from nearshore, sorting and disposal of the light fractions back into the sea. The loss in sediment budget is thought to be compensated by replacement by waves. Net effect of impact of heavy mineral mining on land involves alteration in the surface topography.

Mining of Limestone Deposit

The limestone quarry is located within the Walayar Reserve Forests and mining has led to deforestation in the area. It is estimated that an area of 250 ha. of virgin forest of largely deciduous type has been deforested for various purposes connected with the Malabar Cement Factory. Dumping of waste material into the valley head of one of the tributaries of Cheemantinala which debouches into Malampuzha reservoir, poses threat to the surface water system. Although a dam to check flow of material into the reservoir has been constructed, fines definitely find their way to the reservoir.These problems assume added dimensions in view of the fact that any substantial contamination of the surface water system and large scale transport of waste material through the surface water system shall adversely affect the Malampuzha dam, whose water is used for irrigation in Palghat district and for the water supply scheme of the Palghat town. Hydrochemical data during post-monsoon period indicate increased concentration levels of elements with the cavalues exceeding the permissible limit of water quality (Ramachandran *et al.*, (1989). Constant monitoring of water pollution and sediment transport parameters due to mining there, and strict enforcement of forest protection rules in areas adjoining the quarry are very essential for minimising the damage due to mining there. Air pollution due to dust is more at the cement factory site than at the mine. Dust from the cement factory is flown more than 8-10 km from the factory site at Walayar. Noise pollution at the mining site resulting from blasting drives away the animals, especially elephants, whose pastures are limited in the area.

Glass sand is mined for use of M/s Ogale Glass Co. and for foundry application in Coimbatore and elsewhere. Mining alters the surface morphology and removal of the scanty surface layer.

Clays are the largest mineral resources of Kerala and clay mining industry employees the maximum number of persons in the field of mining though the clay mining centres are generally small in size. Degradation of land due to alteration of surface morphology and removal of surface soil, loss of vegetation etc. are the environmental impacts of clay mining. It is possible to fill the clay mines with the waste material and to partly compensate for the environmental damage. However, this is not strictly practised in the mining sites.

More disastrous damage to the environment is caused by the ubiquitous excavation of tile clays from the paddy fields and river banks brick making. The fine subsurface clay layer from the paddy fields is excavated for brick making, often leaving the paddy fields as large sinks. These sinks accelerate surface soil erosion from the adjoining areas and serve as discharge grounds for groundwater, drying the adjacent high grounds. During periods of inundation, these are partly filled with brackish water and recharge the local ground water system. There are eighteen clay mining sites along the bank of Vamanapuram river alone, causing severe bank failure (Saritha *et al.*, 1996). Considering the large scale excavation of tile clays in Kerala from the paddy fields and the resultant damage to agriculture, vegetation and groundwater regime more attention needs to be paid by government to minimize the hazards.

Limeshell excavation from the backwater leads to depletion of the resource. It also stirs the water column, thereby disturbing the environment necessary for flourishing of new fauna there. Limeshell excavation from land in north Kerala results in the removal of soil, vegetation and hinders cultivation.

The largest, possible area of mining in Kerala pertains to building materials. Building material excavation alters the surface topography, often leaving behind larger pits that get later filled with water. Indiscriminate excavation from mining lease areas, without consideration for other private and public properties often adversely affect the safety of adjacent buildings, roads, and even historical monuments as has been recently reported from the Wayanad region. In hilly terrains, such mining leads to instability of slopes. It also leads to noise and air pollution. Large scale removal of building materials has resulted in obliteration of entire ridges/mounds. Its impact in terms of isostasy needs to be evaluated in the Kerala context.

Illicit mining for chrysoberyl in parts of Thiruvananthpurum district both in the midlands around Nedumangad, Venjaramood etc. and in the forests leads to devastation of vegetation and surface morphological alterations, and leaves large open pits and caves both in forests and in the river beds.

Excavation of sand from river beds leads to deepening of river-beds, accelerating erosion in the upper river course and often leading to lowering of ground water levels in the adjoining areas. Alteration in the river bed morphology also affects the flow characteristics of rivers. The alarming nature of sand mining in Kerala can be gauged from data of a relatively small river basin like Vamanapuram in Kerala. It is noticed that there are 34 mining sites in this basin with an area of 787 km^2, and a channel length of 81 km only for its main course (Saritha *et al.*, 1996). Out of the 34 sites, 6 are situated on the river bank itself. The latter is stated to cause abnormal increase in suspended load.

Conclusion

The foregoing discussion would clearly show that even the limited mining activities in Kerala have already led to substantial environmental degradation, mostly, affecting paddy cultivation, vegetation, river water systems and surface morphology.

As developmental needs call for increased mining activities, considerations of minimising the environmental hazards associated with them and of part preservation of these resources for the posterity should prompt the government to promulgate more effective mineral conservation rules and to undertake impact studies through its existing agencies.

Part damage to the environment by mining can be overcome by filling the mine pits by waste material and this also may not be achieved without intervention by government. A constructive public opinion about the hazards due to mining and about the ways and means of minimising them shall yet be a more effective tool in this direction.

As mining leads to new job opportunities and overall well-being of the population, its socio-economic impact is positive in a state like Kerala, where ethnic and cultural differences are negligible. However, its environmental impact is mostly detrimental.

REFERENCES

Ramachandran, K.K.; Balasubramonian, G. and Bhava Sivadas, 1989. Some Environmental Impact Studies of Limestone Mining in Walayar Area, Palghat District, Kerala. (Final project report submitted to the Kerala State Committee on Science, Technology & Environment), Centre for Earth Sci. Studies, p. 33.

Saritha, D.; Sajikumar, S. and Terry Machado, 1996. Sand and Clay Mining in the Vamanapuram River Basin and Its Environmental Impacts. *Proc VIII Kerala Sci. Congress*, 1995, Kerala State Committee on Science, Technology & Environment, p. 20.

Sivadas, K.M. and Dhruva Rao, B.K., 1975. Exploration Report on the Pandarettu limestone deposit in Walayar forest area, Palghat District, Kerala. *Proc. Symp. on Mineral Resources of Kerala and their Utilization*, Thiruvananthapuram, pp. 90-112.

21

ENVIRONMENT CRISIS, POPULATION GROWTH AND OVERCONSUMPTION IN THE PHILIPPINES
A Development Threat

ELNORA R. PULMA

Introduction

DEVELOPMENT is a global issue. It has always been the vision of all countries in the world, that strategies and measures were conceptualized and implemented to realize such vision. However, development is viewed in different aspects by different theorists and economists that the true concept of development existed for the past three decades. It has evolved from a mere concept of economic growth to something that looks into the whole social system beyond GNP. Such is the case because much of the economic growth in the recent past is not sustainable due to environmental degradation. The concept then extended to a process called sustainable development which does not only look into the present situation but also take into consideration the benefits of the future generation.

Sustainable development then became the global thrust in the 1990s. Philippines for that matter had put an aggressive step towards the attainment of its vision for Philippines 2000. Various policies were formulated and numerous legislations were passed and approved for implementation to respond to the call for a sustainable development. This initiated a total overturn of the country's development strategies. From a mere economic growth strategies which had not brought about great impact in 1970s to sustainable development strategies in 1990s as embodied in the Medium-Term Development Plan of the country for the year 1993-1998.

The Philippines support the contention of most demographers and environmentalists that there is a correlated linkage between environment and population growth. It can never be denied that the current environmental crisis can also be attributed to the world's rapid population growth. This might not be the sole cause but the interrelationship between the two factors of development plays an important role in the whole gamut of development process.

The country has made an early start in the implementation of programmes that seek integrated and holistic development. But challenges remain. The huge population of poor continues to be the major challenge. It has inadvertently brought about an environment that would not be responsive to their needs due to their own unconscious deeds. An indication of hope was felt when the Philippine government started to acknowledged the dismal situation and has tried to responsibly acted on it. How effective and responsive such strategies being implemented still remains to be seen in the future.

It is then the purpose of this chapter to bring about the salient features of interrelationships of environment degradation, population growth and overconsumption in the Philippine context. Analysis on the strategies adopted by the country to respond to the challenge is an added feature of this chapter.

Population Growth and Development

Population growth and development are complex, interlinked processes whereby each affect the other in many ways, and both are interrelated to broad sets of economic, social and political factors.

In the demographic perspective, we can say that the population in many developing countries including the Philippines, will continue to grow up to the latter half of the next century irrespective of fertility trends, although statistics shows that it is already exhibiting a downward trend. In fact, from a rate of 5.24 in 1970, the total fertility rate of the country declined to 4.09 in 1993. Such is the case because the rate at which the population grows depend upon the rate of decline in fertility.

In the particular context of the Philippines, our richness in natural resources is endangered by the continuing rapid population growth, which is exerting pressure to this wealth.This situation also brought about a widespread prevalence of poverty which only widen the gap for the country to attain its desired national development. It has already been a widely known fact that rapid population growth has great implications to any country's progress and it is therefore imperative at this point of time to determine the answers and solutions to the problem that we are now confronting. At this point, implications of population growth shall be dealt with in a sectoral manner so as to emphasize certain points.

A. Population Growth and Renewable Resources (With Attention to Agricultural Land)

Perusal of literature and studies conducted revealed that unlike exhaustible resources (e.g., mineral deposits), natural resources like agricultural land, forest and fisheries can be renewed by natural processes. This means, natural resources are potentially capable of providing economic services for a long time, in fact, forever, as long as their regenerative capacity is not damaged.

We all know that natural resources are input to production. As population grows, there is a greater demand for such inputs in order to produce goods and services needed by the growing population.

A first adap'ive response to the population pressure on the land is to use land extensively, i.e., open new lands for cultivation, as what happened in the Philippines. Output then of the country expands. However, certain problems beset the country because while the land remain fixed, the population grows further which in effect declines the productivity output per worker. This is the case because with population growth, per capita availability of land declines which consequently decline the average output.

Acknowledging this problem, the Philippine government responded to this by changing and adapting new agricultural technology into the agricultural system of the country. The decline in productivity was therefore offset by the introduction of agricultural innovations. More intensive land use has been introduced through the use of irrigation, and application of fertilizers, pesticides as well as using high yielding seed varieties. The introduction of these innovations, for sometime, arrest or reverse the decline in output per worker.

Agricultural innovations, in the country, however did not occur automatically with population growth although population pressure prodded the need for such innovations. Being a country which is predominantly catholic, the objective of the Philippines to reduce its fertility rate finds difficulty in achieving. Conflict between the state and the church continue to exist that population growth outpace the growth of economy and the pace of technological innovations, which only aggravate the whole poverty situation of the country. Moreover, it takes time for these innovations to be introduced. Proper institutions such as research and credit had been developed to introduce these innovation which necessitates a substantial share in the country's national budget.

The strain put into the Philippines' agricultural expenditure is so much that the government in itself cannot respond to the country's requirements. In fact, if we look into some basic statistics of the country, we can see that technology advancement along this aspect had lagged behind other developing countries. The momentum that it had experienced in agricultural sector way back in 1960s was not sustained primarily because of the rapid outburst of population growth in the middle part of 1970s until the later part of the 80s. Table 21.1 presented below would show the insufficiency of agricultural extensions needed by the country to increase its productivity.

TABLE 21.1: Status of land classification and irrigated lands: 1960-1990
(*in percentage*)

Classification	*1960*	*1970*	*1980*	*1990*
Alienable and disposable (arable)	39.61	41.90	44.23	47.06
Forest land	55.30	42.87	36.45	33.50
Others	5.09	15.23	19.32	19.44
Irrigated land (in % of arrable land)	35.20	23.10	20.60	17.20

Source: Philippines Statistical Yearbook, 1991.

Philippines has a total land area of approximately 300,000 km^2 of which as of 1990, 47.06 per cent is considered to be arable land, while 33.50 per cent composed the woodland and forest cover . The remaining 19.4 per cent is comprised of the different land classification types of the country. The dismal fact is that, out of the 47.06 per cent of arable land, only 17.2 per cent are irrigated which of course has great impact to the country's agricultural productivity. It is a known fact that Philippines is one of the tropical countries, with a longer hot climate than the rest of its neighbouring countries. It therefore needs sufficient amount of water for agricultural operation thus, the need for irrigation extension. Rain-feed farming is basically out of the picture taking into consideration the climatic condition of the country. It is just sad to say that while the government is aware of this situation, it cannot allocate sufficient fund for irrigation due to the pressure of basic social services requirements of the rapidly growing population, particularly in younger age group. It is also very apparent that encroachment to forest land is getting larger in scale. As we see in the table, the arable land has extended itself up to the marginal lands of the country and even eat up the forest covers just to satisfy the growing needs of an increasing population either for subsistence or commercial purposes.

It must be noted that population pressure also caused land fragmentation in the Philippines which is also characterized by many developing countries. To the extent that the land fragmentation was not matched by the introduction of intensive and environmentally sustainable agricultural techniques (e.g., intercropping), the farmers with exceedingly low plots are forced to over-use their land or to migrate and engaged in ecologically destructive practices of land extensification on marginal lands where soil and climatic conditions are poorly suited for annual cropping.

Cropland extension into tropical rain forests, has also been a major form of land extensification in the Philippines. Further perusal of existing data revealed that about 143,000 ha. of the total forest and woodland were already deforested for the period between 1980-1989 or in a matter of approximately ten years. The annual rate of deforestation was reported to be about 1.5 per cent during the same period (see Table 21.2).

TABLE 21.2: Selected natural resources indicators: 1980-1990

Indicator	*1980*	*1990*
Forest area (thousand sq. km.)	110,000	78,000
Production of fuelwood and charcoal (1,000 m^3 per year)	26,320	35,038
Deforestation (1,000 ha. per year) 1980-1989		143,000
Annual rate of deforestation (%) 1980-1989		1.5

Source: Human Resource Development, 1995.

Surveys conducted revealed that one of the principal causes of deforestation in the Philippines is the heavy influx of upland migrants who engaged in unsustainable agricultural techniques and shifting cultivation. This is further aggravated by the growing number of population who invaded the forest for fuelwood and charcoal. Table 21.2 would also show the increasing trend of production along this two items.

However, population growth is not the sole culprit of environmental degradation. Certain economic development has also brought about negative environmental consequences particularly in the developing countries, just like the Philippines. The structural adjustment introduced into the economic system of the country through trade liberalization has led negative environmental effects. This can be attested by few studies conducted on this issue.

The studies made in the country revealed that industrial countries just like the United States benefit from the agricultural trade liberalization introduced in the country in both economic and environmental terms. A concrete example was the United States sugar policy. It denied marketing opportunities to efficient developing producers, such as the Philippines, crippling the country's sugar industries and inducing poverty-stricken workers to become slash and burn farmers in the upper watershed which in effect created a harmful consequences to Philippines' natural resources, particularly its forest covers.

Higher agricultural prices internationally also stimulate increased agricultural production in the Philippines through more intensive cultivation which place additional pressure on the land and encourage the expansion of cultivation unto marginal lands or forests.

Based on these given facts, we can say that environmental degradation is actually caused by different factors which are interrelated in various aspects that one cannot ignore its importance in the whole analysis of national development.

B. Population Growth and Environmental Degradation

It has been an accepted fact that the quality of the natural or physical environment (air, water, climate conditions, species of plants and animals) has direct significance on health and production.

Environmental resources i.e., public lands, forest and marine resources are common property resources, and as such there is an inherent tendency for these resources to be overexploited.

As population grows rapidly, which is very much apparent in the Philippine scenario, the demand for its environmental resources also expands since these resources are inputs to production.

Given the collective ownership or common property rights of these resources, rapid population growth in the country has hasten the degradation of these resources.

Degradation of the Philippine environment takes many forms such as: deforestation, soil erosion, deterioration of water resources, destruction of coral reefs, and pollution of the air and water from industrial production. This has really a great negative impact in the development process that various strategies were conceptualized and implemented to arrest environmental degradation. Institutional changes defining property rights were made; private distortions in the market were corrected; and technological innovation in the conservation and use of resources was introduced.

However, these institutional market and technological changes did not occur automatically. They entail costs and time to introduce. In this case, the population of the Philippines grow rapidly, from 27.08 million in 1960 to 60.7 million in 1990, even faster than the capacity to introduce the changes, that environmental resources of the country was at the risk of being irreparably damage, thereby seriously affected the country's future productive capacity. Again, the interlink between population growth and environmental degradation became very apparent in this context.

Destruction of Tropical Forests

It must be noted that there is a great impact of population growth on migration to marginal areas particularly to upland and coastal areas which lead to environmental degradation.

Historical trends of upland population in the Philippines showed that it is continuously increasing for the past thirty (30) years. Fig. 21.1 revealed that from a level of 6.7 million in 1960, upland population in the Philippines had gone up to 17.5 million in 1990, almost about 29 per cent of the total population of the country. Such trend has a great implication on the degradation of natural resources particularly the country's forest covers which suffered the most. This is also one of the contributory factors of low agricultural productivity of the country.

It is true that deforestation was associated primarily with such factors as logging, access roads and trails and expansion of commercial agriculture. However, population growth has been said to be a major factor since 1950. It is apparent that the unprecedented growth of the country's population precipitated migration to upland areas. Migrants from lowland areas pushed the minority cultivators deeper into steeper slopes where they engaged shifting cultivation and non-renewable agricultural techniques, leading to extensive deforestation followed by soil erosion which in effect contributed to scarcity of water supply for agricultural production especially during dry season. This situation ultimately resulted to decreased agricultural productivity. A schematic diagram (see Fig. 21.2) would show that one of the factors attributable to the poor performance of the agricultural sector of the Philippines, is the degraded forest resources brought about by increasing level of migrants to upland areas.

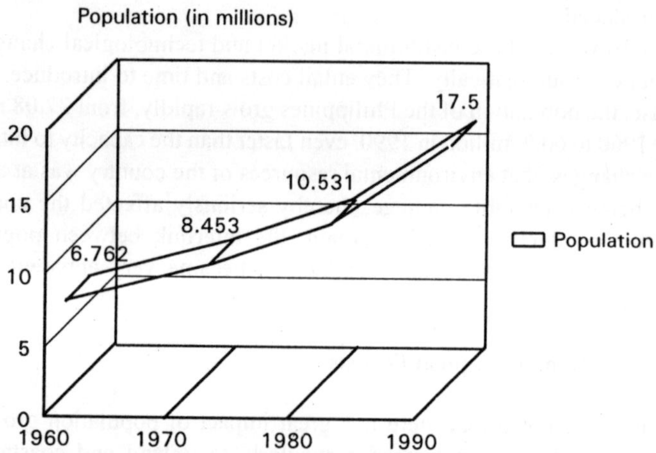

FIG. 21.1: Historical trends of upland population in the Philippines, 1960-1990

Source: Philippines Development Report 1991.

The movement of this people can be attributed to the denseness of population in the lowlands. People, particularly the landless farmers tends to migrate to the upland areas and expand agriculture up the steeper mountain slopes, leading to extensive deforestation followed by soil erosion. By law in the Philippines, all areas with slopes over 18 degrees are considered "public lands", but given lack of effective control over their use, this means they are essentially "open access" lands. Although commercial logging continuous to play an important and destructive role, the conversion of forests to agriculture from land extensification has been the major factor in recent decades.

Cropland expansion then occurs at the expense of forests. If we go by statistics, we can see that because of accelerated migration into forested lands in the Philippines, cultivated area in the uplands grew from 9 per cent of the total cultivated area in the country in 1960 to over 31 per cent in 1987, nearly half of all upland areas are currently cultivated. This increase corresponds to a trebling of the upland population in

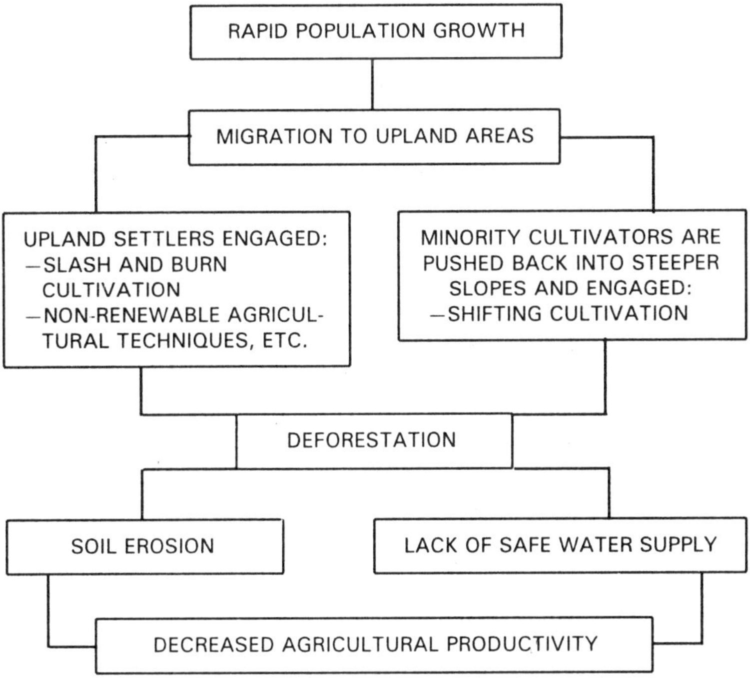

FIG. 21.2: Framework of interrelationship between rapid upland population growth and forest degradation

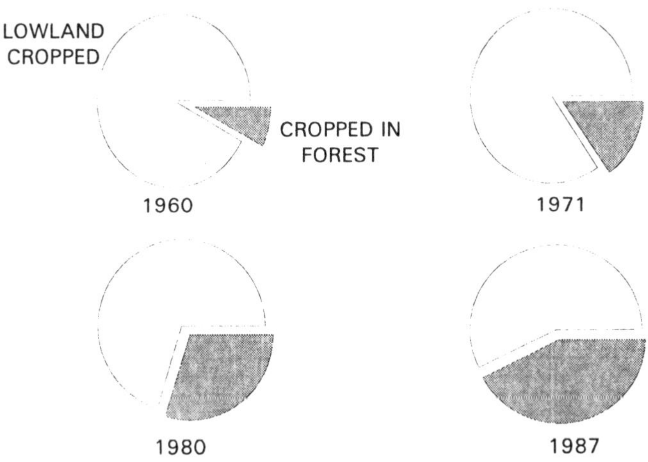

FIG. 21.3: Lowland and forest cropped area, the Philippines, 1960-1987

1950-1985, at an average annual rate of 3.3 per cent, even greater than the country average. Further perusal of existing data revealed that the expansion of cropped area in forests occurred at an annual rate of about 7.6 per cent during 1980-1987 while lowland croplands increased at a rate of only 2 per cent (see Fig. 21.3).

Although cultivable, these forested sloping lands require adequate soil conservation to reduce soil erosion rates and to minimize fertile soil loss. But most of the upland cultivation is destructive for two reasons: First, migrants use land-clearing and land-preparation techniques that tend to be time and labour saving such as "slash and burn" which encourages leaching and soil loss. Secondly, commercial value crops, such as cacao and coffee, are often planted to augment household incomes. However, studies revealed that these crops consume large amounts of soil nutrients, reducing the fertility of the soil after only two years of cropping (World Bank, 1989). In addition, some migrant households combine farming with grazing which also contributed to excessive soil erosion.

The statistics presented leads us to make a conclusive statement that the underlying causes of deforestation in the upland areas of the country can be identified as follows: (a) population growth; (b) commercial logging; (c) extreme land concentration in the lowlands (a few very large farms and many small ones); (d) urban bias in development policies wherein forest policy encourages forest exploitation for contruction of commercial establishments and road networks for industrialization purposes, and etc.

It is within this context, that we can say it is not necessarily appropriate to generalize that environmental trends are only marginally influenced by population growth. Rather, it is the interaction between demographic and socio-economic trends in depletion of natural resource and degradation of the environment in the rural areas of the Philippines.

Degradation of Freshwater Resources

Another basic natural resource of the Philippines that is at the risk of degradation is the country's freshwater. Geographically speaking, Philippines is located in the dry climate tropics and water was becoming increasingly scarce. This can be attributed to various factors but population growth is considered to be the primary cause of this problem. This human-induced water scarcity in the country is therefore being studied so that proper measures can be adapted to minimize and control its occurrence.

Water is a finite resource, but the population of the Philippines was growing rapidly which gives pressure to the finite water resource. In fact, water pollution was a growing problem in the country. This degraded water quality affected human health and welfare directly, while it indirectly contributed to slower rates of development. It is also worth noting that water-related land degradation which is apparent in the Philippines, threatened the sustainable use of resources, in as much as the country is to some extent, still dependent upon irrigated agriculture.

Studies conducted revealed that water scarcity in the Philippines was due to the desiccation of the landscape, a result of large-scale deforestation, destruction of watershed and overgrazing of grasslands. This of course, does not preclude the fact that the increasing demand of rapidly growing population as well as the gross inefficient use of such resource are two contributing factors of the existing problem.

Table 21.3 presented below would clearly show the rate at which freshwater resource of the country has been increasingly degraded over the last 20 years. As we can see, out of the total freshwater resource posted at 29.5 cu km., about 9.1 per cent are withdrawn annually for the period 1970-1992. Furthermore, it is also reported that out of the total freshwater resource withdrawal, 562 cu.m. was for industrial and agricultural purposes, while only about 123 cu.m. was withdrawn for domestic purposes. This data only confirms the contention of many environmentalists that water scarcity is highly attributed to human economic activities.

TABLE 21.3: Freshwater resources: Annual withdrawals of some selected countries, 1970-1992

Country	*Internal Renewable Water Resources (cu. km.)*	*Annual Withdrawal (% of total)*	*Annual withdrawal per capita (cu.m.)*		
			Total	*Domestic*	*Industrial & Agriculture*
China	28.9	7.7	416	54	362
India	380.0	18.2	612	18	594
Philippines	29.5	9.1	686	123	562
Egypt	56.4	97.1	1028	72	956
United Arab Emirates	0.9	299.0	884	97	787
Vietnam	28.9	7.7	416	54	362

Source: World Development Report, 1995.

It has been a recognized fact that water scarcity was one of the ultimate threats to sustainable development and livelihood security. In the long run, rapid population growth of the Philippines aggravated the problem of water scarcity because there is less water per capita to support human life, to improve health and food security and to increase market-oriented activities, such as cash crops and industrial production.

The situation then calls for a revised government plan of action that would focus its effort towards appropriate management of water, land and forest resources of the country as an aggressive step towards the attainment of the country's development goal.

Conclusion and Recommendation

The Philippines today is still beset by so many problems that its goal for sustainable development still remains to be uncertain. The country has to exert extra effort to surpass all these unsurmountable challenges before it can fully propel itself to its desired development.

Rapid population growth, maldistribution of its population, poverty and environmental degradation are factors that are all interrelated in a complex manner. Human activities, such as farming marginal lands, especially in slopes, felling trees for firewood and clearing forest land for agriculture, have degraded the environment and depleted the natural resources. These activities arise from the needs of growing population for land to cultivate subsistence crops. These are just one of the many indicators of population, environment and consumption interrelationships.

Further perusal of documents and statistics led as to a conclusive statement that there exist a linkage between the country's environmental crisis and population growth. The population growth rate of the Philippines, which is still high by international standard, is really putting pressure to the country's limited natural resources.

The vision of Philippines 2000 which is achieving sustainable development, is therefore posing a great challenge to the country. The present situation brings doubt as to what extent can the country do to attain its development goal.

It is therefore imperative at this point of time, for all sectors in the society, to be responsive enough to the call of nature. There is a need to care for the environment not only for the present generation but also for the betterment of the future generation.

The two factors of demography and environment must then be properly responded to. National policies of the country along population and environment should be refocused and redefined along the intricate relationships of demographic factors and socio-economic processes as a strategy to attain a kind of sustainable development that is attuned to Philippine context.

REFERENCES

United Nations, 1994. *Population, Environment and Development*, United Nations Publication.

Republic of the Philippines, 1993. *Philippine Development Report 1987-1992: Environment and Natural Resources.*

Republic of the Philippines, 1992. *Philippine Statistical Yearbook* 1991.

UNDP,1995. Human Development Report 1995. Oxford University Press, Inc.

World Development Report 1995: Workers in an Integrating World, Oxford University Press, Inc.

22

POPULATION AND DEVELOPMENT
Some Issues—India's Profile

B.K. ROY

Introduction

THIS chapter on population and development deliberates on selected issues and patterns of the demographic state interlining, to some extent, the prevalent evaluatory skills and designs so far projected to assess the regional disparities. The paper outlines population characteristics in Section I and status of development in Section II.

Section I—Population Growth: Basic Issues

India's population according to the 1991 census, crossed the 844 million mark with an average annual growth rate of 2.1 per cent. Some states, comparatively backward, have shown high annual growth rates, Andhra Pradesh, Assam and Madhya Pradesh are in this group. West Bengal and Maharashtra have also shown tendencies of high annual growth rate. Average annual growth rate is the lowest (1.4%) in Tamil Nadu and the maximum (2.5%) is in Rajasthan. However, it is important to note that these figures require post enumeration check of 1991 and necessary correction need to be incorporated for a better perception of demographic situation.

The other issue is the frequent changes in the jurisdiction of districts and lower administrative levels from census to census every decade. In spite of freezing the boundaries at various levels the change is quite sharp in the eve of the census. This hampers the activities of preparing grounds to assimilate demographic changes precisely and to find out historical and comparable data from one census to the other. As a result, it is not always possible to compare population in synchronous terms.

Even, immediately after the conduct of the census, several changes in the administrative levels take place and so, the census data become redundant to measure growth in such areas later, because of the lack of base line data. The spirit and purpose of the census are defeated in such cases.

However, geographically, it is seen that number of districts showing decline in growth rate vary from state to state. In all, some 260 districts in major 15 states of the country have recorded decline in growth rate during 1981-91. It is paradoxical that urban sprawling in 390 urban agglomerations in 1991 have increased. And the rural population is rated as urban population due to this change of areal jurisdiction. Large as well as small urban agglomerations in the major states are emerging at the cost of rural space in districts. This has considerable impact on apparent decline in rural population. The decline in total growth is from 2.22 per cent in 1981 to 2.11 per cent in 1991, It may not be primarily due to the reasons of family planning alone, rather to a larger extent, it may be an outcome of redistribution of population, partly internal migration and partly due to conversion and declaration of rural areas as parts of urban space around towns and cities. In general, numerical strength of total population during 1981-91 stands added by > 160 million heads. The urban population of 57.7 Million was added over the 1981 (159.4 Million) amounting to 36.19 per cent of decennial growth. The rural population came down to the mark of 74.28 per cent in 1991 against 76.66 per cent of 1981.

Urban Development and Changes

Urban population has been increasing althrough. Growth is higher in urban areas compared to the rural sector.

A precise picture can emerge after migration data of 1991 is evaluated incorporating jurisdictional changes of rural and urban areas which have taken place during census of 1981 and 1991. It is noted that a large number of small and medium size settlements (803) which were declared as town during the 1951 census, faced declassification in 1961. The reason for this was that these towns had a poor economic base and virtually no infrastructure were created to give impetus to their development. During 1981-91, 63 centers were declassified.

The distribution of urban population during the 1971-81 decade may be examined to assess changes. The districts with higher urban population align in a definite zone due to infrastructural availabilities. It is noteworthy that the districts with high percentages of urban population (> 40%) contributed to a high degree of regional and national development. Also, 12 districts, 4 of which were entirely urban, accounted for greater than 60 per cent urbanization. The next level 40-60 per cent covered 20 districts, which are mostly concentrated in Maharashtra, Uttar Pradesh, Gujarat, Madhya Pradesh and Tamil Nadu. It emerges that higher percentage of urban population was concentrated in lower number of districts. The 1991 situation in the level of urbanization recorded marginal change than the previous decade.

Concomitant to the country wide distribution of urban population the change in urban population during 1981-91 is no less interesting. However, the absolute number of urban dwellers has increased by 49.86 million during 1971-81 and 58 millions in 81-91. The number of towns has also been increasing, Net increase from 71 to 81 was 177 towns, while in 1991 it stood at 389 towns. Most of these towns have inadequate amenities for living. It is not prudent to expand the urban base without first creating a reasonable quality of living. Basic amenities such as drinking water, electricity, job-generation and alike, all needed for a healthy townscape, are absent. During delineation of urban spaces, a great deal of dilution takes place. Due to the typical Indian urban system, expansion of urban area in most cases violates statutory laws and helps expansion of urban area at the expense of local jurisdictional limits. The Indian census projects towns, corporations, municipalities, urban agglomeration, OG (Outgrowth) C.T. (Census towns) and other general systems in the same frame. Many of these areas have few amenities to offer to their population. Services are not created in the same proportion as population increase in such towns. There are many socio economic and political reasons behind such a situation. The census concept of defining a place as urban based purely on demographic parameters has resulted in creation of the census town (CT), or an outgrowth adjunct to an urban place.

The 1981 census recognized 486 outgrowths of the existing cities and towns, while, in 1991, this number has increased to 707. Because of this, 1.58 million rural people were defined as urban dwellers in 1981 and 3.36 million in 1991. Of course, in certain places well planned areas particularly project towns, do have proper town like environments. In certain cases OGs are formed covering rural areas, which are subsequently controlled by adjacent local bodies or project authorities.

The Mega Cities

The 12 mega cities of India, Calcutta, Greater Bombay, Delhi, Madras, Bangalore, Ahmedabad, Hyderabad, Pune, Kanpur, Nagpur, Jaipur and Lucknow accounted for 26.71 per cent of the total urban population of India (42.12 million). Calcutta and Bombay together share 10 per cent of this. To ease the pain of living in such densely populated areas, the Urbanization Commission, set up by the Ministry of Urban Development in 1989, has worked to streamline city orientations and neighbourhoods. It will, however, take some time before urban stress is significantly reduced. It is important to understand that much of the congestion in these urban places is due to rural urban migration, which brings additional stress on resources, employment and other opportunities. The dependency in these urban areas has also increased dramatically. In 1971, the dependency ratio was 1:1.07. In 1981, it increased to 1:1.14. This means that non workers in the urban sector rose from 31.57 Million to 47.768 million, if marginal workers are added to this, the total is then 48.75 million. It is suffice to say that in 1991, the situation has not changed much. At present (as per 1991 census) 49.96 per cent males and merely 9.73 per cent females participate in working force under urban sector.

One of the major factors for growth of mega city has been search for employment by the immigrants. This factor alone accounts for 38.82 per cent (1.64 million) of migrants to Bombay. In Calcutta this accounted for 27 per cent of migrants or 0.77 million people followed by Delhi's 30 per cent or 0.77 million. Family migration in Bombay and Delhi is the highest, with over a million persons in each. It is quite obvious that these cities will over-grow in the future. In addition, 11 other urban centers have reached million city status by 1991. It is anticipated that the existing mega cities will experience substantial population increase by the turn of the century. It is also expected that 16 cities will experience around 100 per cent growth in population by 2001 A.D., keeping the base line population of 1981. Further, it is noted that India will have 36 cities with million population by the turn of the century. A war-footing strategy is required to provide amenities, infrastructure and job opportunities in them. It is quite certain that this conglomeration of population may definitely usher new dimensions in existing urban problems of the country. Threats are already perceived because the urge to develop such centers are critically lacking. It is estimated that Delhi's population will increase from 5.73 million in 1981 to 12 million in 2001; an increase of over 125 per cent. Condition at other cities are no less precarious. It is noteworthy that Calcutta may remain non dynamic even by 2001 AD in spite of the fact that about 2 million population will be added in coming 10 years which is proportionately smaller compared to other million cities. Cities must prepare for the future rise in population. The rural urban migration should be cut short by strengthening rural development and creating urban corridors to face the challenges of the near future.

Urban Corridors—A Development of Urban Macro-space

Urbanization and urban space in India will continue to grow into the 21st century. The reorientation strategies of the National Urbanization Commission may prove fruitful, provided they are formalized in action plans. The concept of urban corridors was initiated by Roy (1977) in an effort to understand correctly the population issues of this country. The urban corridor is a geographical concept and evolved from several factors. These factors include land levels, distribution of urban places of various orders, ecological conditions around the urban places, means of communication, inter-state influences of urban places and potential population. The urban corridors are worked out for 24 axes of cities and towns. India needs to reduce the attraction of flow of population to the cities and mega cities by creating infrastructure within defined corridors and existing cities and streamlining local bodies, so that migrant pressure is redistributed and development of population and such regions are initiated. This concept should be organized by the Central Government so that maximum benefits will be created in particular regions, regional imbalances are addressed, and inter state migrations for employment are reduced. Outgrowths (OGS) Census towns and other settlements should be reorganized to emphasis economic activities which in turn will provide better living standards, more opportunities and a non traditional new order of urban administrative systems in future.

Work Scenario

A direct measure of population and development is the proportion of population engaged in work. Although, it is not an indication of employment, yet it gives a major clue to economic under growth. There is a lot of agreements and disagreements among people about the concept of workers as canvassed in the Census of India. However, the ratio between workers and population in various areas may enlighten issues towards an economic order. The population and worker ratio has been unfavourable in the country and the dependency has been always acute giving rise to unemployment and work participation below any standard in the economic activity. The academic communities and also the government are more concerned about the definition of work rather than to create jobs in rural and urban areas.

Indian society gives preference to males in work due to various reasons. Female workers are confined to households, agriculture and underdeveloped household based industries, the prime reason is being the gaps in their literacy. Taking into account the concept of the main workers during the 1981 and 1991 census, the worker-population ratios both in rural and urban have indicated marginal increase. However, it is also noted that the female workers-population ratio, including both main and marginal workers enumerated by the 1991 census remain significantly lower than those based on principle and subsidiary workers surveyed by the NSS in its large quinquennial surveys of employment and unemployment since 1977-78 (Visaria, 1991). Marginal workers in urban India has remained more or less static at 1 per cent level in 1981-91, while in rural, it has increased from 7.2 per cent in 1981 to 8 per cent in 1991. This indicates increase in unemployment. It is important to note that agricultural workers and cultivators together showed female preponderance over males at both census years of 1981 and 1991.

Section II—Status of Development

The crisis faced by population growth, acute housing problems, depletion of water resources, drinking water shortages, additional creation of agricultural lands, ecological imbalances (forest loss), and alike including mismanagement of resources and skills lead to accumulation of problems from plan to plan.

The evaluatory models of disparities assessment have shown clearly the areas and regions which lagged behind in the development process. Historically, development centered around those districts which were having urban cores of significant growth. These areas attracted investments. Mitra (1964) assessed 34 districts in the top level of the development with another 34 peripheral districts in second level. In all 127 districts accounted for 57.06 per cent of urban population creating some degree of development.

Combining rural, urban and spatial systems of population and dependency on agriculture at the district level as many as 221 districts are found to be lagging behind (depressed) from the point of view of development in India (Roy, 1974). There are 34 districts which can be considered as developed and another 60 districts are transitional. Proportion of non agricultural workers to total population is a simple, but important indicator of development. It is observed that 28 districts in the country record 20 per cent non agricultural population.

Most of them were comparable and coincided to Mitra's and Roy's analysis referred earlier. 335 districts having 5.01 to 20 per cent non-agricultural workers were secondary areas of development. This brought out that the districts showing the development in non agricultural sector and having higher level of urbanization could not be able to create hinterlands, rather, they stood out as islands of development to contribute in the national economy in general. This wide gap created spatial poverty among the majority of the districts of the country.

Thus, it is noted that about 636,000 km^2 area in India is poor with acute ecological depletion and equally 614,000 km^2 is moderately poor with acute poverty where ecological balance is deplorable as well. Only about 675,000 km^2 area in terms of districts (part/full) in the country is developed to some degree which has to share bulk of the burden to carry forward the needs of population and economy of the country. It means regional imbalances are quite pronounced seeking decentralized planning and development. The VIII Five Year Plan (1992-97) formulations have recognized that a third of our population lives in conditions of poverty who are denied of the basic minimum needs. However, geographical dimensions of poverty at regional level is not yet properly recognized.

Spatial Strategies to Regionalization for Development—Agricultural and Demographic Zones

The above patterns of development evaluation lead to the need of creating geographic base of regionalization to consider and study growth patterns within a particular frame to reduce geographic regional poverty and underdevelopment persisting in India. Geography as such has provided important consideration, techniques and methodologies to develop such frames for planning:

- (a) to project provision of basic human needs to target population,
- (b) equitable distribution of benefits of development conceived,
- (c) effective participation of decision making machinery, and
- (d) to create a regular sound data base, both spatial and statistical, to evaluate and overcome gaps in development.

In addition, such regional frames may give a ground of non-changing character of boundaries to overcome the chronic system of creation, merger or bifurcation of districts and other lower level boundaries in the country which hamper greatly the efforts of development and interferes in the data base for planning. In this connection, two most important efforts, viz., (i) the Agro-climatic regionalization of the planning commission and (ii) the regional divisions of India worked out by the census of India, (Ray, 1990) are important. The former may be implemented for agricultural and associated development and the other may form a sound level for generation and analysis of population parameters to integrate with the plans. However, the concept defined in the system of regional divisions may yield a sound frame and fulfil the need of demographic zones of various orders for assessing geographic regional poverty and population resource planning with defined parameters (Bose 1991).

Concluding Remarks

Relating development to population is a complex issue. A few developed districts in the country are still contributing to growth in economy. Regional approach in poverty alleviation can help in a long way to overcome the present problems. The latest trend in population systems has given an indication of various shifts and declines at microscopic variables. It is time that regional systems are considered to augment development of regions in the country which has been overlooked for last 40 years.

REFERENCES

Mitra, A. 1964. *Levels of Development*, Census of India Manager of Publications, Delhi.

Roy, B.K., 1977. 'Typogrammes of Selected Cities and Distribution of Potential Population', *Deccan Geographer*, 15(2), pp. 231-96.

Roy, B.K., 1974. Depressed Areas and Zonation of Districts to the Set up of Physiogeographical Regions of India for Regional Development, *National Geographical Journal of India*, 22(2), pp. 71-89.

Roy, B.K., 1990. Regional Divisions of India—A Cartographic Analysis, Occasional Papers, Series 1 (30 volumes), Office of Registrar General, India, New Delhi.

Visaria, P., 1991. India's Population problems: Emerging perspective after the 1991 Census, *Demography India*, 20(2), pp. 273-95.

INDEX